W9-DGL-542

CLUSTER ANALYSIS

CLUSTER ANALYSIS

Third edition

Brian S Everitt

Professor of Statistics in Behavioural Science, Institute of Psychiatry, London.

A member of the Hodder Headline Group
LONDON • SYDNEY • AUCKLAND

Copublished in the Americas by Halsted Press
an imprint of John Wiley & Sons Inc.
New York — Toronto

To Rachel Morwenna

First published in Great Britain by Heinemann Educational for the
Social Science Research Council 1974
Second edition 1980

Third edition 1993 published by Edward Arnold
Reprinted 1995 by Arnold, a member of the
Hodder Headline Group, 338 Euston Road, London NW1 3BH

Copublished in the Americas by Halsted Press, an imprint
of John Wiley & Sons Inc., 605 Third Avenue, New York,
NY 10158

© 1993 B. S. Everitt

British Library Cataloguing in Publication Data
A catalogue record for this book is available
from the British Library.

ISBN 0 340 58479 3

Library of Congress Cataloging-in-Publication Data
A catalog record for this book is available
from the Library of Congress.

ISBN 0 470 22043 0

Whilst the advice and information in this book is believed to be true and accurate
at the date of going to press, neither the author nor the publisher can accept
any legal responsibility for any errors or omissions that may be made.

Printed and bound in Great Britain by J W Arrowsmith Ltd, Bristol
and Bookcraft, Avon.

Preface

Cluster analysis has been employed in a remarkable number of different disciplines. In psychiatry the techniques have been used to redefine existing diagnostic categories. In archaeology clustering has been used to investigate the relationship between various types of artifact. In market research methods of cluster analysis have been applied to produce groups of consumers with different buying patterns and in anthropology, classifications of the puberty rites of American Indian tribes have been constructed from the results of a cluster analysis.

Clearly workers in a variety of areas need to be aware of the methods of cluster analysis and their possibilities and limitations. It is hoped that, like the previous two editions, this text will provide a convenient and readable introduction to clustering for such workers. In this edition more emphasis is given to examples of the use of clustering methods in addition to describing more recent work. Since many potential users of these techniques will have only limited mathematical or statistical training, the mathematical level of the book has been kept deliberately low, although a certain amount of mathematical nomenclature is unavoidable.

In the first edition of this book I acknowledged the help of a colleague in inserting punctuation according to the rules of the English language rather than by random allocation. In this edition this task has been performed by Dr Graham Dunn, who also commented on the text in other respects. I am very grateful for his support. An 'anonymous' reviewer also made many helpful points which have considerably improved the text. Again I am very grateful for his or her help. Finally thanks are due to Ms Marie Dyer for typing the manuscript.

B.S.Everitt

Institute of Psychiatry, London, 1992

Contents

1

An Introduction to Classification and Clustering

1.1 Introduction

One of the most basic abilities of living creatures involves the grouping of similar objects to produce a classification. Early man for example, must have been able to realize that many individual objects shared certain properties such as being edible, or poisonous, or ferocious, and so on. The idea of sorting similar things into categories is clearly a primitive one since classification, in the widest sense, is needed for the development of language which consists of words which help us to recognize and discuss the different types of events, objects and people we encounter. Each noun in a language, for example, is a label used to describe a class of things which have striking features in common. So animals are named as cats, dogs, horses etc., and such a name collects individuals into groups. *Naming is classifying.*

As well as being a basic human conceptual activity, classification is also fundamental to most branches of science. In biology for example, classification of organisms has been a preoccupation since the very first biological investigations. Aristotle built up an elaborate system for classifying the species of the animal kingdom, which began by dividing animals into two main groups, those having red blood (corresponding roughly to our own vertebrates), and those lacking it (the invertebrates). He further subdivided these two groups according to the way in which the young are produced, whether alive, in eggs, as pupae and so on.

Following Aristotle, Theophrastos wrote the first fundamental accounts of the structure and classification of plants. The resulting books were so fully documented, so profound and so all-embracing in their scope that they provided the groundwork of biological research for many centuries. They were superseded only in the 17th and 18th centuries, when the great European explorers, by opening the rest of the world to inquiring travellers, created the occasion for a second, similar programme of research and collection, under

1

the direction of the Swedish naturalist, Linnaeus. In 1737 Linnaeus published his work *Genera Plantarum*, from which the following quotation is taken.

> "All the real knowledge which we possess, depends on methods by which we distinguish the similar from the dissimilar. The greater number of natural distinctions this method comprehends the clearer becomes our idea of things. The more numerous the objects which employ our attention the more difficult it becomes to form such a method and the more necessary.
>
> "For we must not join in the same genus the horse and the swine, though both species had been one hoof'd nor separate in different genera the goat, the reindeer and the elk, tho' they differ in the form of their horns. We ought therefore by attentive and diligent observation to determine the limits of the genera, since they cannot be determined *a priori*. This is the great work, the important labour, for should the genera be confused, all would be confusion."

In biology the theory and practice of classifying organisms is generally known as *taxonomy*. Initially taxonomy in its widest sense was perhaps more of an art than a scientific method, but eventually less subjective techniques were developed largely by Adanson (1727–1806), who is credited by Sokal and Sneath (1963) with the introduction of the *polythetic* type of system into biology, in which classifications are based on many characteristics of the objects being studied, as opposed to *monothetic* systems, which use a single characteristic to produce a classification.

The classification of animals and plants has clearly played an important role in the fields of biology and zoology particularly as a basis for Darwin's theory of evolution. But classification has also played a central role in the development of theories in other fields of science. The classification of the elements in the periodic table for example, produced by Mendeleyev in the 1860s has had a profound impact on the understanding of the structure of the atom. Again in astronomy, the classification of stars into *dwarf* stars and *giant* stars using the Hertsprung–Russell plot of temperature against luminosity (see Figure 1.1), has strongly affected theories of stellar evolution.

1.2   Reasons for classifying

In the widest sense, a classification scheme may represent simply a convenient method for organizing a large set of data so that the retrieval of information may be made more efficiently. Describing patterns of similarity and differences among the objects under investigation by means of their class labels may provide a very convenient summary of the data. In market research for example, it may be useful to group a large number of respondents according to their needs in a particular product area.

In many applications however, a classification to serve more fundamental purposes may be sought. Medicine provides a good example. Diseases which look the same can often have different causes. A brain haemorrhage is obviously different from an attack of meningitis, although clinically they may appear similar. Conversely, diseases from the same cause can appear utterly remote from one another; for example, rheumatoid arthritis and a particular type of anaemia. To understand and treat disease it has to be classified and the

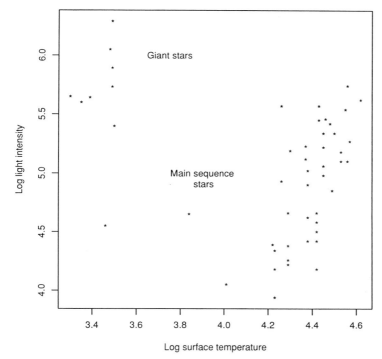

Figure 1.1 Hertsprung–Russell plot of temperature against luminosity.

classification will have two main aims. The first will be *prediction*—separating diseases that require different treatments, the second will be to provide a basis for research into *aetiology*—the causes of different types of disease.

The two aims may not necessarily lead to the same classification and a variety of alternative classifications for the same set of objects or individuals will always exist. Human beings for example, may be classified with respect to economic status into groups such as *lower class, middle class* and *upper class*, or they might be classified by annual consumption of alcohol into *low, medium* and *high*. Obviously different classifications may not collect particular individuals into the same groups. Some classifications will however, be more useful than others, a point clearly made by Needham (1965) when discussing the classification of human beings into men and women.

> "The usefulness of this classification does not begin and end with all that can, in one sense, be strictly inferred from it—namely a statement about sexual organs. It is a very useful classification because classing a person as man or woman conveys a great deal more information, about probable relative size, strength, certain types of dexterity and so on. When we say that persons in class *man* are more suitable than persons in class *woman* for certain tasks and conversely, we are only incidentally making a remark about sex, our primary concern being with strength, endurance etc. The point is that we have been able to use a classification

of persons which conveys information on many properties. On the contrary a classification of persons into those with hair on their forearms between $\frac{3}{16}$ and $\frac{1}{4}$ inch long and those without, though it may serve some particular use, is certainly of no general use, for imputing membership in the former class to a person conveys information in this property alone. Put another way, there are no known properties which divide up a set of people in a similar manner."

In a similar vein a classification of books based on subject matter into classes such as dictionaries, novels, biographies etc., is of wider use than one based on say the colour of a book's binding, since the former is clearly likely to indicate more of the characteristics of a book than the latter.

The important point suggested by both these examples is that any classification is a division of the objects or individuals into groups based on a set of rules—*it is neither true nor false* (unlike say a theory) and should be judged largely on the usefulness of the results.

1.3   Numerical methods of classification

Numerical techniques for devising classifications originated largely in the natural sciences such as biology and zoology in an effort to rid taxonomy of its traditionally subjective nature and to provide *objective* and *stable* classifications. Objective in the sense that the analysis of the same set of organisms by the same sequence of numerical methods will produce the same classification; stable in that the classification remains the same under a wide variety of additions of organisms or of new characters. Whether these criteria are always met by the numerical methods will be the subject of discussion in later chapters.

The second half of the twentieth century has seen a dramatic increase in the number of numerical classification techniques available. This growth has largely paralleled the development of high-speed computers, such machines being needed to undertake the large amounts of arithmetic generally involved. As well as an increase in the variety of numerical classification methods, a similar expansion has taken place in the areas of their application. Nowadays such techniques are used in fields as disparate as archaeology and psychiatry, and market research and astronomy.

A number of names have been applied to these methods depending largely on the area of application. *Numerical taxonomy* is generally used in biology. In psychology the term *Q analysis* is sometimes employed. In the artificial intelligence literature *unsupervised pattern recognition* is common. In other areas *clumping* and *grouping* have also been used occasionally. Nowadays however, the most common generic term is *cluster analysis* and it is this term which is largely used in the remainder of this text. The problem which these techniques address may be stated broadly as follows:

> Given a collection of *n* objects individuals, animals, plants etc., each of which is described by a set of *p* characteristics or variables, derive a useful division into a number of classes. Both the number of classes and the properties of the classes are to be determined.

The solution generally sought is a *partition* of the *n* objects, that is a set of clusters where an object belongs to one cluster only, and the complete set

Table 1.1   Small set of multivariate data

Individual	Sex	Age (yrs)	IQ	Depression	Weight (pounds)
1	Male	21	120	Yes	150
2	Male	43	N.K.	No	160
3	Female	22	135	No	135
4	Male	86	150	No	140
5	Female	28	120	Yes	150
6	Female	40	100	Yes	N.K.

N.K. = Not known.

of clusters contains all the objects. (It is important to bear in mind however, that the conclusion of a classification investigation may well be that any such summarisation of the data would be misleading).

The basic data for cluster analysis is, in general, a matrix $\mathbf{X}$, giving the variable values for each of the objects or individuals under investigation; that is,

$$\mathbf{X} = \begin{pmatrix} x_{11} & x_{12} & \cdots & x_{1p} \\ x_{21} & x_{22} & \cdots & x_{2p} \\ \vdots & & & \vdots \\ x_{n1} & x_{n2} & \cdots & x_{np} \end{pmatrix}$$

A small example of such a matrix is given in Table 1.1. Note that the variables may be of different types, a point which will be taken up in detail later. Additionally some values may be missing, a topic to be discussed in Chapter 3.

The aim of clustering the data as described above is to group or cluster the individuals or objects (or whatever—individuals or objects will be the terms used throughout the text), represented by the n rows of $\mathbf{X}$. There appears, however, to be no reason why the clustering should not be applied to $\mathbf{X}'$ to obtain a classification of the variables which describe each object. The distinction between 'object' and 'variable' is perhaps less clear cut than previously implied. Nevertheless in much of the remainder of this text interest will largely centre on the clustering and classification of the rows of $\mathbf{X}$. It should be remembered however that many of the clustering techniques to be described (but not perhaps all) could be applied to the clustering of variables. Indeed such an exercise might, at times, be quite helpful as an alternative to the often uncritical use of techniques such as factor analysis. (More discussion of the clustering of variables is given in Chapter 7.)

Another point to make explicit here is that in this text the main concern is with classifying previously unclassified material, i.e. at the start of the investigation the number and composition of the classes is unknown. It is this that differentiates clustering as a technique for the analysis of multivariate data from *discrimination* and *assignment* methods, where groups are known *a priori*. Such techniques will be encountered only peripherally in subsequent chapters. A detailed account is available in Hand (1981).

1.4   What is a cluster?

Up to this point the terms cluster, group and class have been used in an essentially intuitive manner without any attempt at formal definition. In fact it turns out that such formal definition is not only difficult but may even be misplaced. Bonner (1964), for example, has suggested that the ultimate criterion for evaluating the meaning of such terms is the value judgement of the user. If using a term such as 'cluster' produces an answer of value to the investigator that is all that is required. Other authors, for example Cormack (1971) and Gordon (1980), attempt to define a cluster using properties such as *internal cohesion* and *external isolation*. Such properties are illustrated most simply by considering examples where the number of variables observed on each individual or object is two, so that the data may be plotted directly as a *scattergram*. Examples of clusters with internal cohesion and/or external isolation are shown in Figure 1.2. The 'clusters' present will be clear to most observers without attempting an explicit explanation or definition. Indeed the example indicates that no single definition is likely to be sufficient for all situations.

It is not entirely clear how a 'cluster' is recognised when displayed in the plane, but one feature of the recognition process must involve the assessment of the relative *distances* between points. Consequently the measurement of distance is likely to be an important consideration in any automatic, numerical procedure which attempts to mimic in higher dimensions the eye-brain system for identifying clusters in two-dimensional data. This is a topic which will be taken up in Chapter 3.

A further set of two-dimensional data is plotted in Figure 1.3. Here most observers would conclude that there is no cluster structure; simply a single collection of points. Ideally then, one might expect a method of cluster analysis applied to such data to come to a similar conclusion. As will be seen later this may not be the case, and many methods of cluster analysis *will* divide the type of data seen in Figure 1.3 into 'groups'. Often the process of dividing a homogeneous data set into different parts is referred to as *dissection*, and such a procedure may be useful in specific circumstances. If, for example, the points in Figure 1.3 represented the geographical locations of houses in a town, dissection might be a useful way of dividing the town up into compact postal districts which contain comparable numbers of houses—see Figure 1.4. (This example was suggested by Gordon, 1980). The problem is, of course, that since in most cases the investigator does not know *a priori* the structure of the data, there is a danger of interpreting clustering solutions in terms of the existence of distinct clusters even when in fact this is not the case. This is a very real problem in the application of clustering techniques and one which will be the subject of further discussion in later chapters.

Figure 1.2 Clusters with internal cohesion and/or external isolation. (Reproduced with permission of Chapman and Hall from Gordon, 1980).

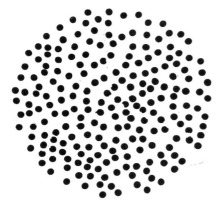

Figure 1.3 Data containing no cluster structure. (Reproduced with permission of Chapman and Hall from Gordon, 1980).

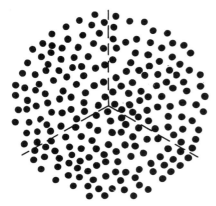

Figure 1.4 Dissection of data in Figure 1.3. (Reproduced with permission of Chapman and Hall from Gordon, 1980).

1.5   Examples of the use of clustering

To illustrate the range of disciplines in which cluster analysis has been applied a number of examples will now be briefly described. Fuller accounts of some of these will appear later at appropriate points in the text.

1.5.1   Psychiatry

Diseases of the mind are more elusive than diseases of the body and there has been much interest in psychiatry in using cluster analysis techniques to refine or even redefine current diagnostic categories. Much of this work has involved depressed patients where interest primarily centres on the question of the existence of *endogenous* and *neurotic* subtypes. Pilowsky *et al.* (1969) for example, clustered 200 patients on the basis of their responses to a depression questionnaire, together with information about their mental state, sex, age and length of illness. (Notice once again the different types of variable involved.) One of the clusters produced was identified with endogenous depression. A similar study by Paykel (1971) using 165 patients and a clustering method due to Friedman and Rubin (1967) (see Chapter 5), indicated four groups one of which was clearly psychotic depression. A review of the classification of depression is given in Farmer *et al.* (1983).

Cluster analysis has also been used to find a classification of individuals who attempt suicide, which might form the basis for studies into the causes and treatment of the problem. Paykel and Rassaby (1978) for example, studied 236 suicide attempters presenting at the main emergency service of a city in the USA. From the pool of available variables, 14 were selected as particularly relevant to classification and used in the analysis. These included age, number of previous suicide attempts, severity of depression and hostility, plus a number of demographic characteristics. A number of cluster methods were applied to the data and a classification with three groups was considered the most useful. The general characteristics of the groups found were as follows:

Group 1:
Patients take overdoses, on the whole showing less risk to life, less psychiatric disturbance, and more evidence of interpersonal rather than self-destructive motivation.

Group 2:
Patients in this group made more severe attempts, with more self-destructive motivation, by more violent methods than overdoses.

Group 3:
Patients in this group had a previous history of many attempts and gestures, their recent attempt was relatively mild, and they were overly hostile, engendering reciprocal hostility in the psychiatrist treating them.

A further application of cluster analysis to parasuicide is described in Kurtz *et al.* (1987).

1.5.2   Medicine

Wastell and Gray (1987) describe a fascinating use of clustering for the development of a classification in patients with *temporomandibular pain dysfunction syndrome.* The main interest here was to develop an objective typology for classifying facial pain in terms of its spatial distribution, with the hope that clinically, the derived classification might be useful in identifying different stages of the disease. This would be of help in defining more directed treatment plans. More details of this study are given in Chapter 4.

1.5.3   Social services

Jolliffe *et al.* (1982) give an account of the use of clustering to derive a classification of the elderly which might be helpful in suggesting how social services could be economically and effectively allocated, and which might also identify groups with particular requirements or which are particularly isolated. Here over 2000 individuals described by 35 characteristics were available for analysis. These were divided prior to clustering proper into those over, and those under 75 years of age. Samples of 600 individuals from each age group were then subjected to cluster analysis.

1.5.4   Market research

A large number of cities are available that could be used as test markets, but due to economic factors testing must be restricted to only a small number of these. Clustering the cities into a small number of groups such that cities within a group are very similar to each other, and then choosing one city from each group, could be used as means of selecting the test markets. Green *et al.* (1967) adopted this approach, classifying 88 cities on the basis of 14 variables such as city size, newspaper circulation, per capita income and so on.

1.5.5   Education

Aitkin, Anderson and Hinde (1981) clustered teachers into distinct styles on the basis of several binary variables describing teaching behaviour, for example: Do pupils have a choice of where to sit? Is a timetable used for organizing work? Are end of term tests given? Are stars given to pupils who produce the best work? The clusters produced (by a method known as *latent class analysis*, see Chapter 6) were identified as 'formal' and 'informal' styles of teaching.

1.5.6   Archaeology

Hodson(1971) uses a *k means* clustering technique (see Chapter 5), to construct a taxonomy of hand axes found in the British Isles. Variables used to describe each of the axes included length, breadth and pointedness at the tip. The analysis resulted in two clusters one of which contained thin, small axes and the other thick, large axes.

1.6   Summary

Techniques of cluster analysis seek to separate a set of data into its constituent groups or clusters. Ideal data for such an analysis would yield clusters so obvious that they could be picked out, at least in small scale cases, without the need for complicated mathematical techniques and without a precise definition of the term 'cluster'. In practice however things are rarely so straightforward, so there has been a great proliferation of clustering techniques over the last three decades or so. Many of these have taken their place alongside other *exploratory* data analysis techniques as tools of the applied statistician. The term exploratory is important here since it explains the largely absent 'p value', ubiquitous in many other areas of statistics, although there have been attempts to set some clustering techniques in an inferential framework as will be seen in later chapters. In the main however, clustering methods are intended largely for *generating* rather than *testing* hypotheses.

2

The Initial Examination of Multivariate Data

2.1 Introduction

According to Jain and Dubes (1988) 'cluster analysis is a tool for exploring data and should be supplemented by techniques for visualizing data'. Certainly it is often useful to begin the search for clusters in multivariate data by examining some relatively simple graphical displays. Well chosen displays can enhance the understanding of the data and may also provide a partial antidote to the dangerous habit of applying clustering techniques in a careless and uncritical way. In this chapter graphical methods designed for an initial exploration of the data are described. In later chapters graphical techniques useful for the display and evaluation of clustering solutions will be described. To begin, it will be assumed that the observed variables are continuous.

2.2 Univariate and bivariate views

It is well known that the marginal distributions of multivariate data do not necessarily reflect accurately their multivariate structure. (An example is given later in this section). Nevertheless it is often very useful to begin the examination of multivariate data by examining distributions on each separate variable by way of *histograms, stem-and-leaf displays, boxplots, non-parametric density estimates* etc. and plots of each pair of variables using a scattergram. Such diagrams may give evidence of patterns or structure in the data, in particular the presence of clusters. Alternatively they may indicate that the data do not contain distinct groups, thus making the application of clustering less compelling. To illustrate what can be achieved by this relatively simple approach the famous Fisher–Anderson iris data will be used. These data are shown in Table 2.1. They consist of measurements of *sepal length, sepal width, petal length* and *petal width* on 150 specimens of iris plant. Three species of iris

11

are involved *Iris setosa, Iris versicolor* and *Iris virginica*, although this labelling is ignored here. To begin, a nonparametric density estimation procedure (see Wegman, 1972), is used to examine the distributions of each of the four variables. Plots of these are shown in Figure 2.1. The estimated densities for petal length and petal width display distinct *multimodality*, which might be taken as preliminary evidence of the presence of clusters in the data.

A further useful 'look' at the data can be obtained from the scattergrams for each pair of variables. These are more easily scanned if they are arranged in *lower triangular* form following the suggestions of Tukey and Tukey (1981) and Chambers *et al.* (1983). The resulting diagram is shown in Figure 2.2. A number of the plots indicate very clear clustering in the data, although it is quite difficult to pick out three groups matching the three different types of iris plant.

It is sometimes helpful to combine the univariate views provided by histograms, box plots etc. with the bivariate scattergrams of the data, on the same diagram. Figure 2.3 shows the iris data displayed in this way. The diagram nicely illustrates the comment made earlier about marginal views of multivariate data not necessarily reflecting their true multivariate structure.

2.3   Accommodating extra variable values on a scattergram

A variety of suggestions have been made as to how further variable values might be included on the basic scatterplot. Gower (1967a), for example, describes a method where other variable values may be accommodated using lines of various lengths emanating from the points in the scattergram. Gnanadesikan (1977) gives an example using circles of various sizes to accommodate a third variable value. Such a diagram is referred to by Wilkinson (1992) as a *bubble plot*. To illustrate this technique it will be applied to the city crime rate data shown in Table 2.2. (These data are taken with permission from Hartigan, 1975). Murder and rape rates provide the scattergram, and autotheft rates the size of the circles. The resulting plot is shown in Figure 2.4. Some points immediately apparent from this plot are that autotheft rates for most cities are similar, but Hartford, Honolulu, Tucson, Portland and Atlanta have rather lower rates. Atlanta is also unusual in having relatively high murder and rape combined with a low rate of autotheft, and Boston is also unusual in this respect but in the reverse direction—it has low murder and rape rates but the highest autotheft rate.

Further techniques for including other variable values on a scattergram are discussed in the next section. All such methods are limited in the number of observations which can be accommodated before the diagram becomes too crowded to be useful.

2.4   Icon representations of multivariate data

When there are only a moderate number of observations, each may often usefully be represented by some type of *symbol* or *icon*, the shape and form of

Table 2.1 Fisher's *Iris* data.

Iris setosa				Iris versicolor				Iris virginica			
Sepal length	Sepal width	Petal length	Petal width	Sepal length	Sepal width	Petal length	Petal width	Sepal length	Sepal width	Petal length	Petal width
5·1	3·5	1·4	0·2	7·0	3·2	4·7	1·4	6·3	3·3	6·0	2·5
4·9	3·0	1·4	0·2	6·4	3·2	4·5	1·5	5·8	2·7	5·1	1·9
4·7	3·2	1·3	0·2	6·9	3·1	4·9	1·5	7·1	3·0	5·9	2·1
4·6	3·1	1·5	0·2	5·5	2·3	4·0	1·3	6·3	2·9	5·6	1·8
5·0	3·6	1·4	0·2	6·5	2·8	4·6	1·5	6·5	3·0	5·8	2·2
5·4	3·9	1·7	0·4	5·7	2·8	4·5	1·3	7·6	3·0	6·6	2·1
4·6	3·4	1·4	0·3	6·3	3·3	4·7	1·6	4·9	2·5	4·5	1·7
5·0	3·4	1·5	0·2	4·9	2·4	3·3	1·0	7·3	2·9	6·3	1·8
4·4	2·9	1·4	0·2	6·6	2·9	4·6	1·3	6·7	2·5	5·8	1·8
4·9	3·1	1·5	0·1	5·2	2·7	3·9	1·4	7·2	3·6	6·1	2·5
5·4	3·7	1·5	0·2	5·0	2·0	3·5	1·0	6·5	3·2	5·1	2·0
4·8	3·4	1·6	0·2	5·9	3·0	4·2	1·5	6·4	2·7	5·3	1·9
4·8	3·0	1·4	0·1	6·0	2·2	4·0	1·0	6·8	3·0	5·5	2·1
4·3	3·0	1·1	0·1	6·1	2·9	4·7	1·4	5·7	2·5	5·0	2·0
5·8	4·0	1·2	0·2	5·6	2·9	3·6	1·3	5·8	2·8	5·1	2·4
5·7	4·4	1·5	0·4	6·7	3·1	4·4	1·4	6·4	3·2	5·3	2·3
5·4	3·9	1·3	0·4	5·6	3·0	4·5	1·5	6·5	3·0	5·5	1·8
5·1	3·5	1·4	0·3	5·8	2·7	4·1	1·0	7·7	3·8	6·7	2·2
5·7	3·8	1·7	0·3	6·2	2·2	4·5	1·5	7·7	2·6	6·9	2·3
5·1	3·8	1·5	0·3	5·6	2·5	3·9	1·1	6·0	2·2	5·0	1·5
5·4	3·4	1·7	0·2	5·9	3·2	4·8	1·8	6·9	3·2	5·7	2·3
5·1	3·7	1·5	0·4	6·1	2·8	4·0	1·3	5·6	2·8	4·9	2·0
4·6	3·6	1·0	0·2	6·3	2·5	4·9	1·5	7·7	2·8	6·7	2·0
5·1	3·3	1·7	0·5	6·1	2·8	4·7	1·2	6·3	2·7	4·9	1·8
4·8	3·4	1·9	0·2	6·4	2·9	4·3	1·3	6·7	3·3	5·7	2·1
5·0	3·0	1·6	0·2	6·6	3·0	4·4	1·4	7·2	3·2	6·0	1·8
5·0	3·4	1·6	0·4	6·8	2·8	4·8	1·4	6·2	2·8	4·8	1·8
5·2	3·5	1·5	0·2	6·7	3·0	5·0	1·7	6·1	3·0	4·9	1·8
5·2	3·4	1·4	0·2	6·0	2·9	4·5	1·5	6·4	2·8	5·6	2·1
4·7	3·2	1·6	0·2	5·7	2·6	3·5	1·0	7·2	3·0	5·8	1·6
4·8	3·1	1·6	0·2	5·5	2·4	3·8	1·1	7·4	2·8	6·1	1·9
5·4	3·4	1·5	0·4	5·5	2·4	3·7	1·0	7·9	3·8	6·4	2·0
5·2	4·1	1·5	0·1	5·8	2·7	3·9	1·2	6·4	2·8	5·6	2·2
5·5	4·2	1·4	0·2	6·0	2·7	5·1	1·6	6·3	2·8	5·1	1·5
4·9	3·1	1·5	0·2	5·4	3·0	4·5	1·5	6·1	2·6	5·6	1·4
5·0	3·2	1·2	0·2	6·0	3·4	4·5	1·6	7·7	3·0	6·1	2·3
5·5	3·5	1·3	0·2	6·7	3·1	4·7	1·5	6·3	3·4	5·6	2·4
4·9	3·6	1·4	0·1	6·3	2·3	4·4	1·3	6·4	3·1	5·5	1·8
4·4	3·0	1·3	0·2	5·6	3·0	4·1	1·3	6·0	3·0	4·8	1·8
5·1	3·4	1·5	0·2	5·5	2·5	4·0	1·3	6·9	3·1	5·4	2·1
5·0	3·5	1·3	0·3	5·5	2·6	4·4	1·2	6·7	3·1	5·6	2·4
4·5	2·3	1·3	0·3	6·1	3·0	4·6	1·4	6·9	3·1	5·1	2·3
4·4	3·2	1·3	0·2	5·8	2·6	4·0	1·2	5·8	2·7	5·1	1·9
5·0	3·5	1·6	0·6	5·0	2·3	3·3	1·0	6·8	3·2	5·9	2·3
5·1	3·8	1·9	0·4	5·6	2·7	4·2	1·3	6·7	3·3	5·7	2·5
4·8	3·0	1·4	0·3	5·7	3·0	4·2	1·2	6·7	3·0	5·2	2·3
5·1	3·8	1·6	0·2	5·7	2·9	4·2	1·3	6·3	2·5	5·0	1·9
4·6	3·2	1·4	0·2	6·2	2·9	4·3	1·3	6·5	3·0	5·2	2·0
5·3	3·7	1·5	0·2	5·1	2·5	3·0	1·1	6·2	3·4	5·4	2·3
5·0	3·3	1·4	0·2	5·7	2·8	4·1	1·3	5·9	3·0	5·1	1·8

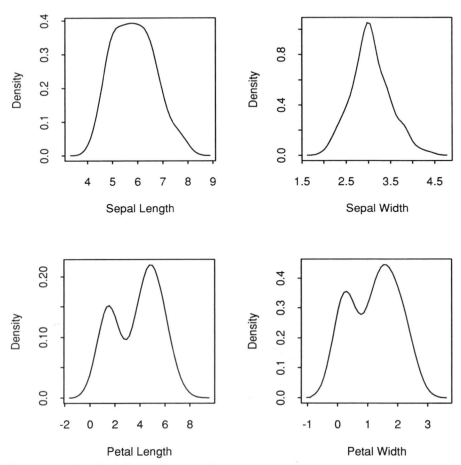

Figure 2.1   Density estimates for variables in iris data.

which is governed by the observation's variable values. Such a representation may be helpful in indicating interesting patterns or structure, in particular the presence or otherwise of clusters of similar observations. There are a number of possibilities:

Profiles,
which represent each observation by p vertical bars, each bar corresponding to a particular variable value and having a height proportional to the value taken by the variable.

Stars or *polygons,*
which represent each variable as a value along equally spaced radii from a common centre. The points of the radii are usually connected.

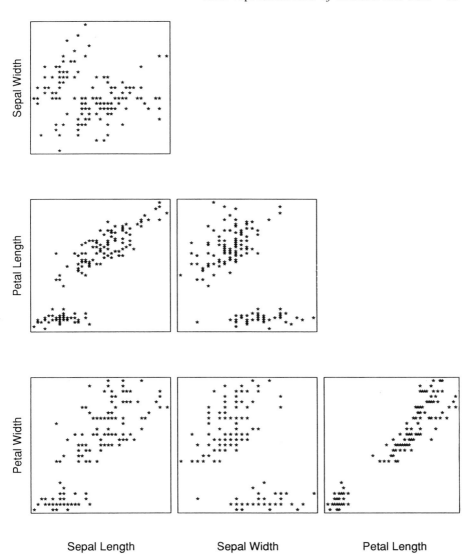

Figure 2.2   Scatterplots for iris data.

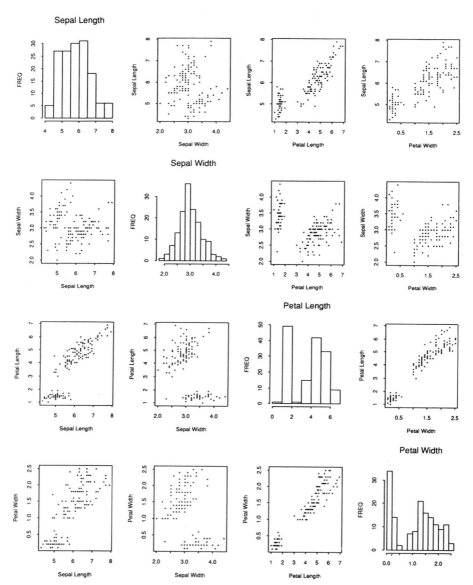

Figure 2.3 Scatterplots and histograms for iris data.

Table 2.2 City crime data (Reproduced with permission from Hartigan, 1975)

	Murder Mansltr.	Rape	Robbery	Assault	Burglary	Larceny	Auto theft
Atlanta (AT)	16·5	24·8	106	147	1112	905	494
Boston (BO)	4·2	13·3	122	90	982	669	954
Chicago (CH)	11·6	24·7	340	242	808	609	645
Dallas (DA)	18·1	34·2	184	293	1668	901	602
Denver (DE)	6·9	41·5	173	191	1534	1368	780
Detroit (DT)	13·0	35·7	477	220	1566	1183	788
Hartford (HA)	2·5	8·8	68	103	1017	724	468
Honolulu (HO)	3·6	12·7	42	28	1457	1102	637
Houston (HS)	16·8	26·6	289	186	1509	787	697
Kansas City (KC)	10·8	43·2	255	226	1494	955	765
Los Angeles (LA)	9·7	51·8	286	355	1902	1386	862
New Orleans (NO)	10·3	39·7	266	283	1056	1036	776
New York (NY)	9·4	19·4	522	267	1674	1392	848
Portland (PO)	5·0	23·0	157	144	1530	1281	488
Tucson (TU)	5·1	22·9	85	148	1206	756	483
Washington (WA)	12·5	27·6	524	217	1496	1003	739

From the United States Statistical Abstract (1970) per 100,000 population.

Faces,
which represent each observation by a cartoon face the features of which—shape, curve of the mouth, position of the eyes etc.—are governed by the observations variable values. The technique was proposed originally by Chernoff (1973) and extended by Flury and Riedwyl (1981).

Trees,
which represent each variable as the length of a branch of a tree whose structure is determined by applying a hierarchical clustering algorithm (see Chapter 4) to the variables. This method is described in detail in Kleiner and Hartigan (1981).

As an illustration, Figure 2.5 shows a collection of bar profiles representing the city crime data, and Figure 2.6 the 'star' representations of the same data. The latter appear to allow similarities and differences of observations to be judged far more effectively than the former. (The single star shown in Figure 2.7 shows how the crime rates for different crimes correspond to particular points of the star).

To illustrate the use of a Chernoff faces representation the data in Table 2.3, taken with permission from Dawkins (1989), will be used. These data consist of the national record times for men for eight running events in 55 countries. The faces representation is shown in Figure 2.8. Readers are encouraged to examine this diagram for evidence of outliers and clusters! This particular way of representing multivariate data has been heavily criticised by some statisticians, their main concern being the degree of subjectivity likely to be involved in assessing similarities between faces.

Direct use of icons has perhaps limited usefulness, but combined with a two-

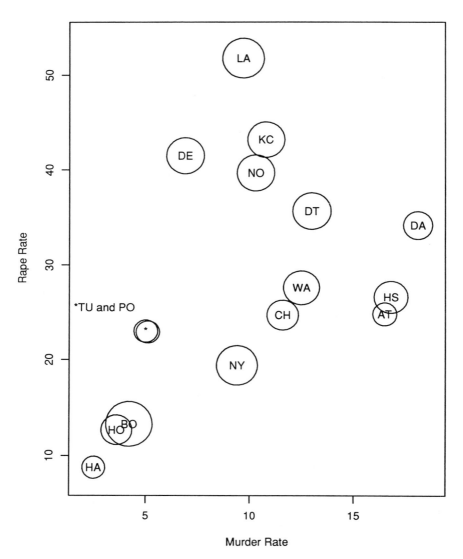

Figure 2.4   Bubble plot for city crime data.

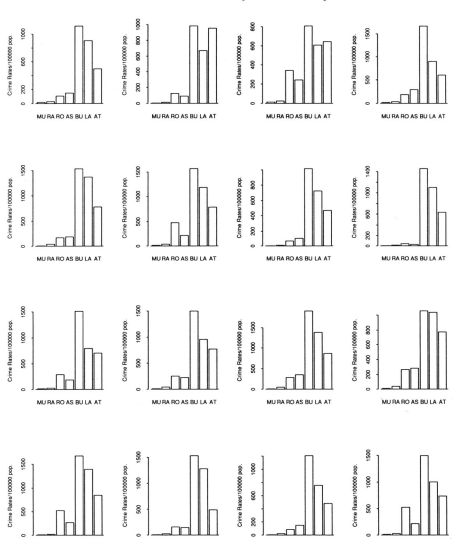

Figure 2.5   Crime rate profiles for USA cities.

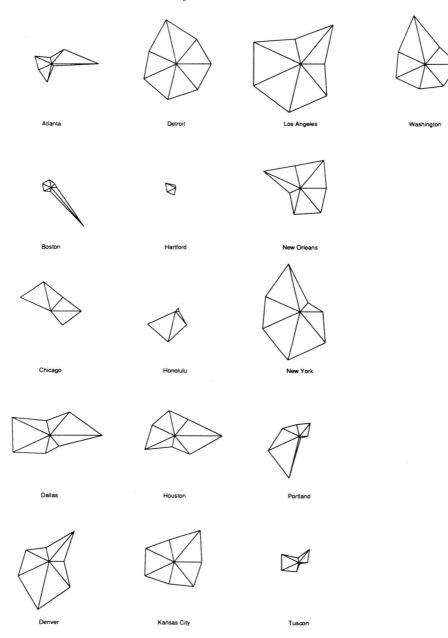

Figure 2.6   Stars representing city crime data.

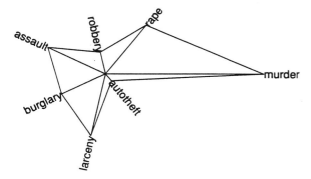

Figure 2.7   Star icon for Atlanta.

dimensional scatterplot they can be very informative. Figure 2.9, for example, consists of a scatterplot of the 100m and 200m times from the athletic data, enhanced by the presence of a six-pointed star at each point indicating the times of the remaining six events. This plot provides a useful summary of the main features of the data. Clearly the athletes of the Cook Islands (CI) are unlikely to be a threat in any of the events considered! The athletes of the Dominican Republic (DR) have relatively good times for the 100m, 200m and 400m but a rather poor time for the marathon. New Zealand (NZ) athletes have relatively poor sprint performances (perhaps the good sprinters are lured at an early age into Rugby Union?), but excel in middle distance and marathon running. The United States (US) and Italy (IT) have the best overall records.

2.5   Andrews' plots

Andrews (1972) proposed the following simple method of plotting multivariate data in two dimensions. Each p-dimensional point $\mathbf{x}' = (x_1, \ldots, x_p)$, where x_i, $i = 1, \ldots, p$ are the variable values, is represented by the function

$$f_{\mathbf{x}}(t) = \frac{1}{\sqrt{2}}x_1 + x_2\sin(t) + x_3\cos(t) + x_4\sin(2t) + x_5\cos(2t) + \ldots$$

plotted over the range $-\pi < t < \pi$. It can be shown that this representation preserves Euclidean distances (see Chapter 3), in the sense that two observations with very similar sets of variable values will be represented by curves that remain close together for all values of t. Two observations whose variable values differ markedly, on the other hand, will be represented by curves which also differ markedly, at least for some values of t. Consequently examination of a set of Andrews' curves may allow the identification of outliers, clusters or other interesting features of the data.

Several Andrews' plots for the same data set can be constructed simply by *permuting* the variables and recalculating the functions $f_{\mathbf{x}_i}(t)$. Since in general, the low frequencies (i.e. those associated with x_1, x_2, x_3) are distinguished more clearly on the plots than the high frequencies (i.e. those associated with x_{p-2},

Table 2.3 Athletic records for 55 countries (Reproduced with permission from Dawkins (1989))

Country	1	2	3	4	5	6	7	8
Argentina	10·39	20·81	46·84	1·81	3·70	14·04	29·36	137·72
Australia	10·31	20·06	44·84	1·74	3·57	13·28	27·66	128·30
Austria	10·44	20·81	46·82	1·79	3·60	13·26	27·72	135·90
Belgium	10·34	20·68	45·04	1·73	3·60	13·22	27·45	129·95
Bermuda	10·28	20·58	45·91	1·80	3·75	14·68	30·55	146·62
Brazil	10·22	20·43	45·21	1·73	3·66	13·62	28·62	133·13
Burma	10·64	21·52	48·30	1·80	3·85	14·45	30·28	139·95
Canada	10·17	20·22	45·68	1·76	3·63	13·55	28·09	130·15
Chile	10·34	20·8	46·20	1·79	3·71	13·61	29·30	134·03
China	10·51	21·04	47·30	1·81	3·73	13·90	29·13	133·53
Colombia	10·43	21·05	46·10	1·82	3·74	13·49	27·88	131·35
Cook Is	12·18	23·2	52·94	2·02	4·24	16·70	35·38	164·70
Costa Rica	10·94	21·9	48·66	1·87	3·84	14·03	28·81	136·58
Czech	10·35	20·65	45·64	1·76	3·58	13·42	28·19	134·32
Denmark	10·56	20·52	45·89	1·78	3·61	13·50	28·11	130·78
Dom Rep	10·14	20·65	46·80	1·82	3·82	14·91	31·45	154·12
Finland	10·43	20·69	45·49	1·74	3·61	13·27	27·52	130·87
France	10·11	20·38	45·28	1·73	3·57	13·34	27·97	132·30
GDR	10·12	20·33	44·87	1·73	3·56	13·17	27·42	129·92
FRG	10·16	20·37	44·50	1·73	3·53	13·21	27·61	132·23
GB	10·11	20·21	44·93	1·70	3·51	13·01	27·51	129·13
Greece	10·22	20·71	46·56	1·78	3·64	14·59	28·45	134·60
Guatemala	10·98	21·82	48·40	1·89	3·80	14·16	30·11	139·33
Hungary	10·26	20·62	46·02	1·77	3·62	13·49	28·44	132·58
India	10·60	21·42	45·73	1·76	3·73	13·77	28·81	131·98
Indonesia	10·59	21·49	47·80	1·84	3·92	14·73	30·79	148·83
Ireland	10·61	20·96	46·30	1·79	3·56	13·32	27·81	132·35
Israel	10·71	21·00	47·80	1·77	3·72	13·66	28·93	137·55
Italy	10·01	19·72	45·26	1·73	3·60	13·23	27·52	131·08
Japan	10·34	20·81	45·86	1·79	3·64	13·41	27·72	128·63
Kenya	10·46	20·66	44·92	1·73	3·55	13·10	27·80	129·75
Korea	10·34	20·89	46·90	1·79	3·77	13·96	29·23	136·25
P Korea	10·91	21·94	47·30	1·85	3·77	14·13	29·67	130·87
Luxemburg	10·35	20·77	47·40	1·82	3·67	13·64	29·08	141·27
Malaysia	10·40	20·92	46·30	1·82	3·80	14·64	31·01	154·10
Mauritius	11·19	22·45	47·70	1·88	3·83	15·06	31·77	152·23
Mexico	10·42	21·30	46·10	1·80	3·65	13·46	27·95	129·20
Netherlands	10·52	29·95	45·10	1·74	3·62	13·36	27·61	129·02
NZ	10·51	20·88	46·10	1·74	3·54	13·21	27·70	128·98
Norway	10·55	21·16	46·71	1·76	3·62	13·34	27·69	131·48
Png	10·96	21·78	47·90	1·90	4·01	14·72	31·36	148·22
Philippines	10·78	21·64	46·24	1·81	3·83	14·74	30·64	145·27
Poland	10·16	20·24	45·36	1·76	3·60	13·29	27·89	131·58
Portugal	10·52	21·17	46·70	1·79	3·62	13·13	27·38	128·65
Rumania	10·41	20·98	45·87	1·76	3·64	13·25	27·67	132·50
Singapore	10·38	21·28	47·40	1·88	3·89	15·11	31·32	157·77
Spain	10·42	20·77	45·98	1·76	3·55	13·31	27·73	131·57
Sweden	10·25	20·61	45·63	1·77	3·61	13·29	27·94	130·63
Switzerland	10·37	20·45	45·78	1·78	3·55	13·22	27·91	131·20
Taipei	10·59	21·29	46·80	1·79	3·77	14·07	30·07	139·27
Thailand	10·39	21·09	47·91	1·83	3·84	15·23	32·56	149·90
Turkey	10·71	21·43	47·60	1·79	3·67	13·56	28·58	131·50
USA	9·93	19·75	43·86	1·73	3·53	13·20	27·43	128·22
USSR	10·07	20·00	44·60	1·75	3·59	13·20	27·53	130·55
W Samoa	10·82	21·86	49·00	2·02	4·24	16·28	34·71	161·83

Event: (1) 100 m (s), (2) 200 m (s), (3) 400 m (s), (4) 800 m (min), (5) 1500 m (min), (6) 5000 m (min), (7) 10000 m (min), (8) Marathon (min).

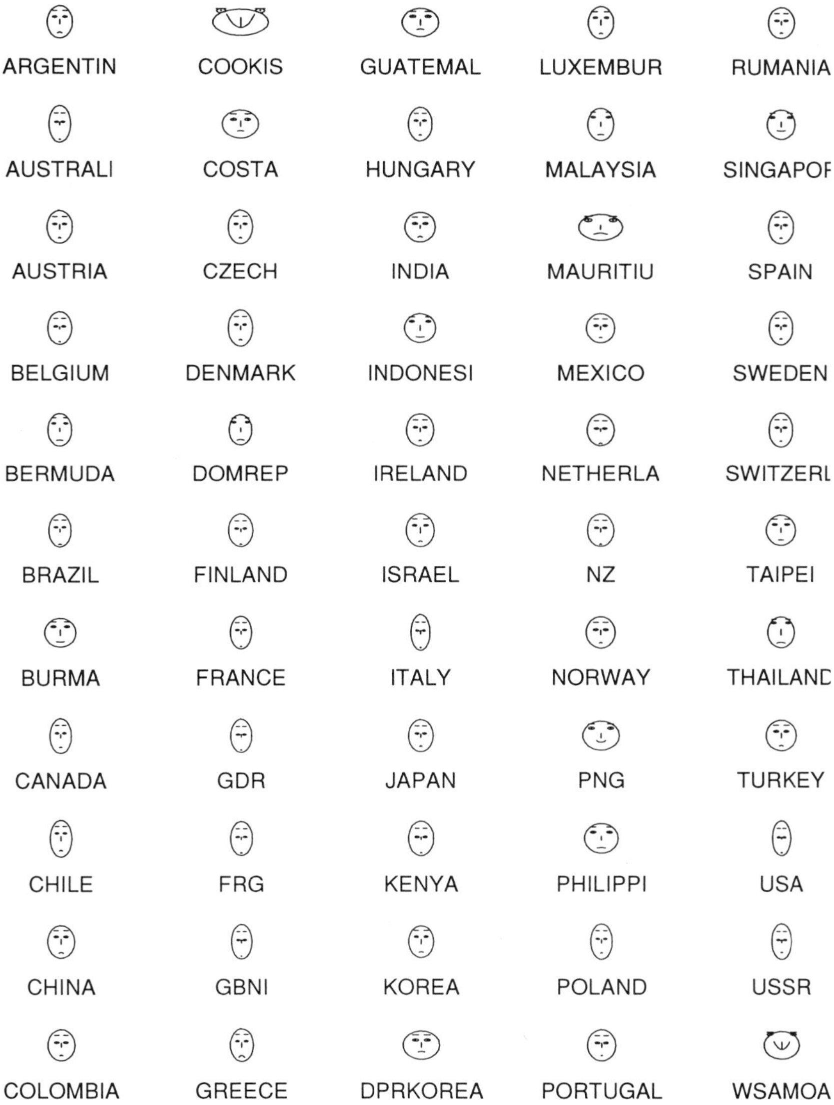

Figure 2.8 Faces representation of athletic records data

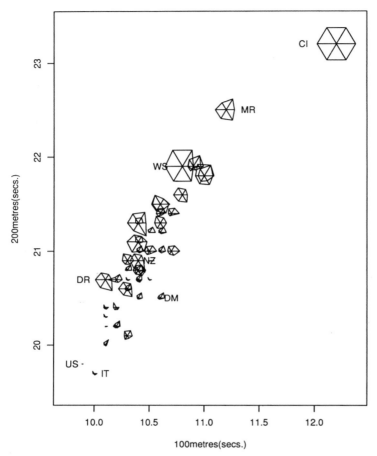

Figure 2.9   Stars plotted on 100m×200m scatterplot of athletic records data

x_{p-1}, x_p), it is best to associate those variables that are thought to be most important for classification with the low frequencies. In practice, of course, this may not be easy, since the investigator may not have a clear *a priori* ordering in such terms. Embrechts and Herzberg (1991) suggest that rather than calculating them from the raw data, Andrews' curves should be obtained from the data after standardization of each variable to zero mean and unit standard deviation. This is said to prevent consistently large variable values in **x** masking visually the effect of other variables in the plotted functions.

To illustrate how this method works in practice, a set of curves were obtained for the iris data, both unstandardized and after the variables had been standardized, according to the suggestion of Embrechts and Herzberg. The resulting plots are shown in Figure 2.10.   In both diagrams a clear

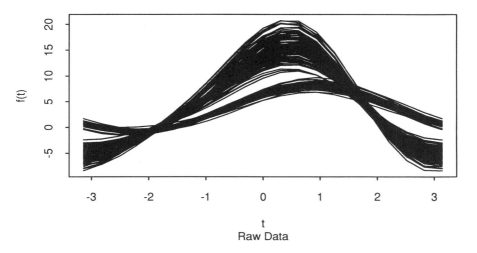

Raw Data

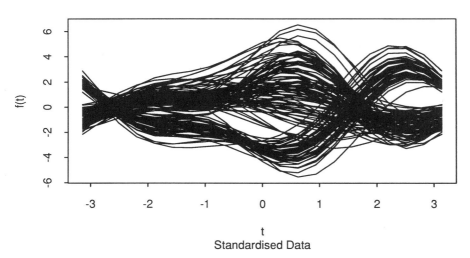

Standardised Data

Figure 2.10 Andrews' plots of iris data.

indication is given of distinct groups of observations, although neither plot has a clear advantage in this respect.

A problem with the use of Andrews' technique in practice is that only a fairly limited number of observations may be plotted on the same diagram. Attempts to plot a large number of curves is very likely to result in confusion rather than clarity. Gnanadesikan (1977) makes some suggestions to deal with this problem.

2.6   The use of derived variables

Many of the techniques described in previous sections may be most useful when applied not to the raw data but to derived scores seeking to summarize the data in some optimal way. There are a number of possible methods of obtaining such scores, *factor analysis, multidimensional scaling, canonical analysis* etc., but perhaps the most common (and often the most useful) method is by means of a *principal components analysis.*

2.6.1   Using principal component scores to display multivariate data

Principal components analysis is described in detail in a number of texts concerned with multivariate analysis, for example Krzanowski (1988) and Everitt and Dunn (1991). It is also the central topic of the monograph by Jolliffe (1986) and the book by Jackson (1991). The essential feature of the technique is the transformation of the original variables $x_1, x_2, \ldots, x_p$ into a new set of variables $y_1, y_2, \ldots, y_p$. The new variables are *linear* transformations of the original with the following properties.

(1) $y_1, y_2, \ldots, y_p$ are uncorrelated with each other.
(2) $y_1, y_2, \ldots, y_p$ account for decreasing portions of the variance of the original variables.

The coefficients defining the linear transformations from $x_1, x_2, \ldots, x_p$ to $y_1, y_2, \ldots, y_p$, are found from the eigenvectors of the covariance matrix, or more commonly the correlation matrix, of the original variables. (In some cases 'labels' may be given to the derived variables, but this is not essential and often not desirable). In cases where the first few of the derived variables, i.e. the *principal components*, account for a large proportion of the variance in $x_1, x_2, \ldots, x_p$, they may be used as a low-dimensional summary of the original data. In particular, pairs of principal component scores can be plotted to allow visual inspection of the structure in the data. (Although in some instances it is possible that low-dimensional principal component plots could conceal the clusters present in a data set—see Figure 2.11.)

As a first example of the use of principal components analysis it was applied to the correlation matrix of the city crime data with the results shown in Table 2.4. Here the first two components account for 68% of the variance in the original data. The first component represents essentially an overall crime rate, the second a contrast between 'violent' and 'non-violent' crimes. A plot of the cities using their first two principal component scores is shown in Figure 2.12, and a plot of all pairs of the first four component scores arranged in lower triangular form in Figure 2.13. Perhaps the most interesting aspect of this figure is the appearance of Boston as an outlier in those plots involving component number three. On this component, Boston has a low score because of its high autotheft rate which here has a negative loading. Variation on this component would decrease considerably if Boston was excluded.

In Figure 2.14 a plot of the first two principal component scores for each city is enhanced by a star representation of their raw data. An interesting feature

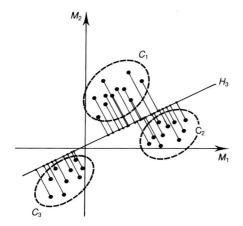

Figure 2.11 A principal components plot which would conceal cluster structure.

Table 2.4 First four principal components of crime rates data

Crime	PC1	PC2	PC3	PC4
Murder	0·53	− 0·70	0·28	− 0·21
Rape	0·81	− 0·07	0·24	0·48
Robbery	0·71	− 0·22	− 0·42	− 0·44
Assault	0·85	− 0·31	0·06	0·18
Burglary	0·72	0·46	0·31	− 0·24
Larceny	0·65	0·68	0·14	− 0·11
Autotheft	0·56	0·13	− 0·71	0·23
Variance	3·45	1·33	0·94	0·63

of this plot is that although Washington and New Orleans have very similar principal component scores, their star representations indicate that their crime profiles are not as similar as might be imagined.

An alternative method for indicating possible distortions produced by a principal components plot is the *minimum spanning tree* which gives the minimum length pathway between the observations such that (a) no closed loops occur, (b) each data point is visited by at least one line and (c) the pathway is connected. Details are given in Gower and Ross (1969). By plotting the links in the minimum spanning tree on to say the plot of the first two component scores, it is often possible to highlight misleading relationships in the diagram. An example will help to illustrate the possibilities. Again the city crime data is used and Figure 2.15 shows these data plotted in the space of the first two principal components with the minimum spanning tree of the Euclidean distances (see Chapter 3) from the original data after standardization, superimposed. The positions of Honolulu and Boston in the plot appear to indicate that they are rather far apart in terms of their crime profiles. The structure of

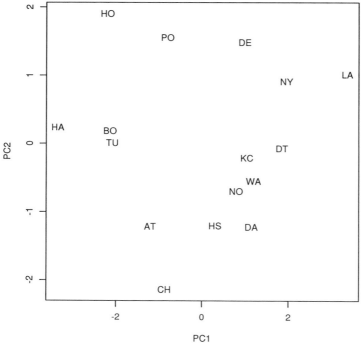

Figure 2.12   Plot of city crime data in space of first two principal components.

the minimum spanning tree however shows this not to be the case, otherwise the two cities would not be joined by one of the links in the tree.

(In the discussion above it has been implicitly assumed that the principal components with the largest eigenvalues will be best for displaying the separation between groups. This is, however, not necessarily always so; see Chang, 1983).

Principal components analysis can be used in association with Andrews' plotting technique to provide an ordering of variables in calculating the curves. The x_1 variable in $f_x(t)$ could now be associated with the first principal component score, x_2 with the second etc. To illustrate this possibility and to compare it with curves produced both from the raw data, and the raw data standardized to zero mean and unit standard deviation, Andrews' plots for the principal component scores of the iris data were found. They are shown in Figure 2.16. Here the plots do not give such convincing evidence of the presence of clusters as the plots in Figure 2.10.

A relatively straightforward extension of the theory behind principal components analysis allows a *joint* display of both the rows and the columns of the multivariate data matrix to be produced. This technique known as the *biplot* is described in detail in Gabriel (1981). Here only an example of its use involving the athletic data is given. The biplot diagram for these data is shown

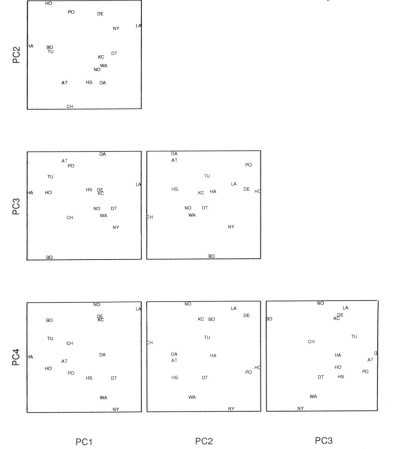

Figure 2.13 Lower triangular plot of first four principal component scores from city crime data.

in Figure 2.17. In this diagram the lengths of the arrows reflect the variances of the corresponding variable, and the angles between them indicate the size of their correlations, small angles corresponding to high correlations. The positions of the points corresponding to the countries, relative to each other, indicate similarities and differences amongst their athletic record profiles. The positions of these points relative to the lines representing variables, reflect a country's times on the various events. Figure 2.17 illustrates nicely the pattern of correlations between the times for the various events. It also highlights the rather low athletic prowess of athletes of the Cook Islands (12) and Western Samoa (55). The Cook Island 400m and 800m runners seem to be particularly poor. Bermuda (5), Malaysia (35), Singapore (46), Thailand (51) and the Dominican Republic (16), appear to form a small 'cluster' of countries whose

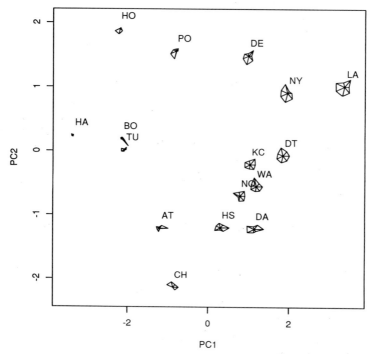

Figure 2.14   Stars plotted on principal component plot of city crime data.

athletes are particularly poor at marathon running. Italy (29) and the USA (53) are seen to have the best overall athletic records.

The use of principal components analysis in association with clustering techniqes will be discussed further in later chapters.

2.6.2   Displaying categorical data using correspondence analysis

All of the data sets used to illustrate the techniques described in the previous sections have involved quantitative variables. In many applications however, the multivariate data of interest will involve categorical, nominal scale variables and it therefore becomes of concern as to how such data may be displayed graphically before more formal and complex methods of clustering are applied. One such technique which is becoming increasingly popular is *correspondence analysis*, a method for visually interpreting multivariate categorical data. Most commonly used is *simple correspondence analysis* which is applied to data described by two categorical variables. Here a graphical display of the corresponding two-way contingency table is produced by deriving coordinate values to represent row and column categories. Like principal components analysis these coordinates are found as the eigenvectors of a particular matrix. Details are given in Everitt and Dunn (1988).

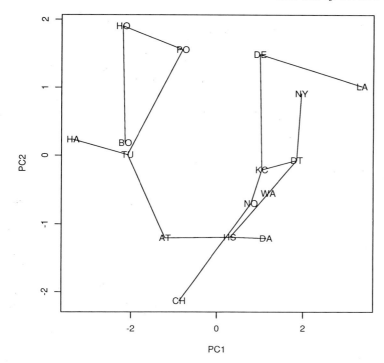

Figure 2.15 Plot of city crime data in space of first two principal components with superimposed minimum spanning tree.

Multiple correspondence analysis is an extension of correspondence analysis to the case of three or more categorical variables. Here graphical displays are produced in which the categories of variables *and* the individual cases are represented as points. The latter can often be examined for evidence of clustering. Details of the technique are given in Greenacre (1984).

Two sets of data will be used to illustrate how correspondence analysis might be used, particularly in association with some of the other graphical techniques described previously. The first of these, given in Table 2.5, shows the distribution of cell sizes in the brains of a number of patients who had died of AIDS and a number of patients who died from other causes. Interest here centres more on whether there is any evidence of differences in the cell-size distributions of the two types of patient rather than clustering *per se*. Nevertheless it provides an interesting example of how correspondence analysis can be used to explore complex categorical data. Regarding the data as a two-dimensional contingency table, simple correspondence analysis may be applied. Here five dimensions are needed to account for over 80% of the *inertia* (essentially the usual chi-squared statistic for testing independence in the table). Pairs of these coordinate values for the 20 individuals have been plotted in lower triangular form to give Figure 2.18. There appears to be

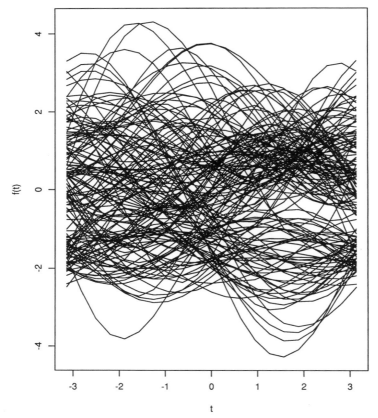

Figure 2.16   Andrews' plots of principal component scores from iris data.

evidence of a number of outliers particularly amongst the AIDS deaths. There seems, however, to be little clear distinction between the AIDS patients and the others.

To illustrate the use of multiple correspondence analysis, a data set involving violent incidents which took place in a secure psychiatric ward will be used. Each such incident was described by a number of categorical variables, such as severity of injury—none, minor, major; time of day—night, morning, afternoon; weapon used—yes, no, etc. One hundred such incidents were thus described and interest centred on discovering a possible typology of violent incidents which might be useful in ward management. Here the first five correspondence coordinates were extracted accounting for 50% of the inertia. They are plotted in Figure 2.19. Very little evidence of clustering is visible in this diagram.

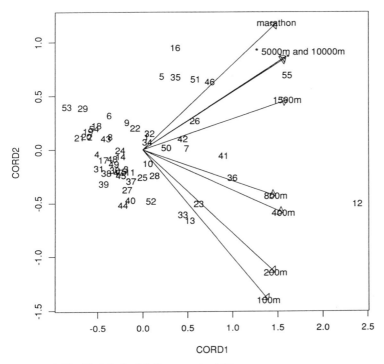

Figure 2.17 Biplot of athletic records data.

2.7 Summary

In this chapter a variety of techniques suitable for providing informative graphical displays of multivariate data have been described. None are specifically
designed for indicating the presence of clusters but they are often useful for
this purpose and, in addition, are often even more useful for preventing exces-
sive claims for cluster structure produced by more complex techniques. This
point will be raised again in later chapters. The techniques can be used on
the raw data or on derived scores such as those from a principal components
or correspondence analysis. In addition, principal component scores can be
used directly to provide two-dimensional views of the data. (Other methods
for projecting the data into a low-dimensional space to allow the structure to
be examined visually are given in Klein and Dubes, 1989).

Table 2.5 Cell size distributions

Patient	s1	s2	s3	s4	s5	s6	s7	s8	s9	s10	s11	s12	s13
						Cell size							
1	8	8	20	10	9	5	3	1	4	1	1	3	0
2	2	11	10	12	5	3	2	2	2	3	0	0	0
3	18	12	10	18	14	10	8	5	1	3	2	2	0
4	19	12	7	11	9	6	3	1	3	1	2	1	0
5	14	17	8	8	2	4	3	3	1	3	1	2	0
6	10	16	9	4	6	4	0	1	2	0	0	0	0
7	3	1	4	3	3	2	2	0	2	0	2	1	1
8	27	22	16	7	6	5	2	2	4	2	0	0	0
9	12	16	16	17	9	7	4	5	1	2	0	0	1
10	10	14	10	10	7	3	1	0	0	2	1	1	0
11	2	3	12	7	4	9	6	2	3	3	2	3	1
12	21	19	9	4	3	2	1	0	0	0	0	0	0
13	20	17	11	7	2	6	1	0	1	0	0	0	0
14	16	27	22	12	9	2	3	3	2	4	2	0	0
15	11	15	13	14	9	9	5	2	2	0	0	1	0
16	10	13	5	12	5	9	5	1	1	2	3	3	0
17	23	13	14	8	4	6	2	3	1	1	2	1	2
18	15	14	15	10	7	12	5	4	0	1	1	2	0
19	14	15	13	11	7	4	1	2	1	0	0	0	0
20	10	19	16	19	8	3	4	1	1	1	1	1	0

The first 11 patients are those dying of AIDS. s1 to s13 represent increasing cell size

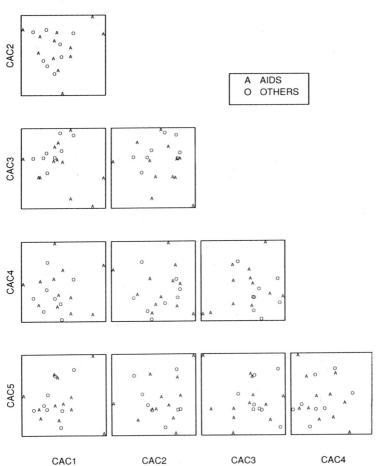

Figure 2.18 Lower triangular plot of coordinates from correspondence analysis of cell size distribution data.

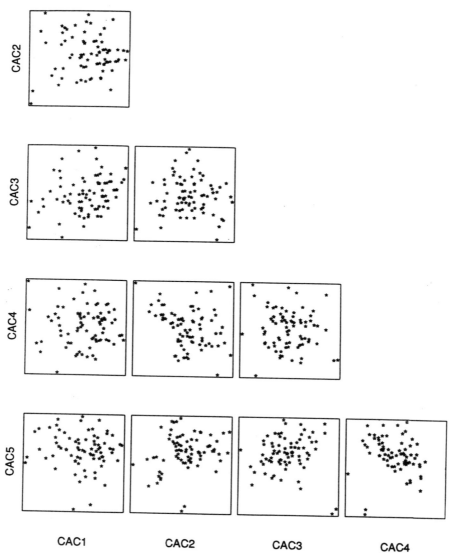

Figure 2.19 Lower triangular plot of coordinates from multiple correspondence analysis of violent incident data.

3

Measurement of Similarity, Dissimilarity and Distance

3.1 Introduction

Many methods of cluster analysis begin not with the raw multivariate data matrix, **X**, met in Chapter 1, but with a matrix containing numbers indicating the *similarity* or the *dissimilarity* of each pair of individuals or objects which are to be clustered. (In the remainder of this chapter the term *proximity* will often be used to refer to both types of index). There are many ways in which proximity may be measured, depending largely on the type of data involved, in particular the type of variable. Many of these methods will be described in this chapter. Before this however, it will be helpful to spend a little time considering some issues surrounding the choice of variables used to describe the objects or individuals to be clustered.

3.2 The choice of variables

The initial choice of the particular set of measurements used to describe each individual to be clustered constitutes a frame of reference within which to establish the clusters; this choice presumably reflects the investigator's judgement of relevance for the purpose of classification. Consequently the first question to ask about the chosen variables is whether they are relevant to the type of classification being sought. For example, if a classification of the mentally ill is the goal, useful for assessing the effects of different treatments, it is probably not sensible to include such variables as height, weight and other vital statistics, since this might cause the resulting clusters to be simply 'males' and 'females'. It is important to bear in mind therefore, that the initial choice of variables is itself a categorization of the data for which there are only limited statistical and mathematical guidelines. (Similar remarks could be made about the individuals or objects chosen for study).

The next question that might be considered is how many variables should be measured on each individual? In most cases there are likely to be a theoretically limitless number of variables which might be used to produce a classification. In practice of course, many will be deemed irrelevant for the purpose at hand, and a further restriction on numbers will arise from temporal or economic considerations. As with the question of which variables to measure, there is, in general, no sound theoretical basis for determining the number of variables to use and the problem must therefore be approached empirically. In most applications it is probable that the researcher will err on the side of taking more rather than less measurements, which can give rise to computational problems with some clustering techniques. More importantly the presence of additional variables on which clusters are not distinguished may obscure the cluster structure. De Sarbo *et al.* (1984) illustrate this problem with the following example. In automobile marketing research applications, it is often found that certain attitudinal variables, such as those emphasizing styling and comfort versus design simplicity and high fuel mileage, produce clusters of car owners with markedly different patterns of brand ownership. If, however, extraneous attitudinal variables are added (describing general leisure-time interests or feature preferences) the original structure may be completely obscured.

A further problem, common to all branches of multivariate analysis, is the possibility of *missing values*. These can occur for a variety of reasons and may be dealt with in a number of ways. The simplest is to consider only individuals who have a complete set of variable values. In some cases however this can severely reduce the number of individuals available for analysis. An alternative approach is to replace the missing values by estimated values. It is often suggested, for example, that missing values be estimated by the mean of the variable in question. For some multivariate techniques this might be a reasonable approach. In cluster analysis it is not. Here the mean should clearly be calculated only from those individuals belonging to the same group as the individual with the incomplete data. But such a group-specific calculation is however not possible because the groups are, of course, unknown. More satisfactory methods for estimating missing values in the context of cluster analysis are described in Dixon (1979) and Little and Rubin (1987).

3.2.1   Standardization

In many applications the variables describing the objects to be clustered will not be measured in the same units. Indeed they may often be variables of completely different types, some *categorical*, others *ordinal*, and yet others having an *interval* scale. (Table 1.1 shows such a data set). It is clear that it would not be sensible to treat say, weight measured in pounds, height measured in inches, and anxiety rated on a four-point scale as equivalent in any sense, in determining a measure of similarity or distance. For interval scaled variables, the solution suggested most often is to simply standardize each to unit variance prior to any analysis, using the standard deviations calculated from the complete set of objects to be clustered. Fleiss and Zubin (1969) however, show that this may have the serious disadvantage of diluting

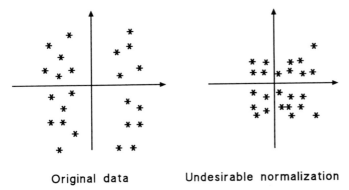

Original data   Undesirable normalization

Figure 3.1 Illustration of standardization problem. (Taken with permission of Prentice Hall from Jain and Dubes, *Algorithms for Clustering Data*, 1988.)

differences between groups on the variables which are the best discriminators, a point also made by Duda and Hart (1973). Figure 3.1 illustrates the problem. Using the within-group standard deviations to standardize would largely overcome this difficulty; in a clustering context however, these are not available. A further disadvantage of standardizing each variable separately is that it ignores possible correlations between the variables. Various ways of allowing for correlated variables when measuring distance or similarity are discussed in later sections of this chapter.

When the variables measured are of different types, various suggestions have been made as to how they might be used together. The simplest approach is to convert all variables into binary form before calculating similarities. The age variable in Table 1.1 could, for example, be transformed into 'below 40' and '40 and above'. This has the advantage of being very straightforward, but the disadvantage of sacrificing potentially useful information. A more attractive alternative is to use a similarity coefficient which can incorporate information from different types of variable in a sensible fashion. Such a coefficient is described in Section 3.3.3. A further possibility for dealing with mixed variable types is to carry out separate analyses of the same set of objects, each analysis involving variables of a single type only, and to attempt to synthesize the results from the different studies. Finally a method of clustering designed specifically for this type of data might be used; see Chapter 6.

3.2.2 Weighting of variables

To weight a variable means to give it greater or lesser importance than other variables when using these to produce a classification. Now although it is perfectly possible to arrange for certain variables to be weighted, various authors question the validity of such a procedure—see for example, Sneath and Sokal (1973), their reason being that the weights can only be based on intuitive judgements of what is important, and that these may simply reflect existing classifications of the data. In general this is not what is required in a clustering application. More commonly the methods of cluster analysis are applied to a data set in the hope that previously unnoticed and potentially useful groups will emerge. Hitherto accepted groups may perhaps not turn up because they are not supported by the present data, and the investigator may be forced to revise hypotheses or propose a better measurement space.

The argument in favour of equal weighting is not however straightforward. In studies where, for example, a subset of all possible variables has been selected, zero weights have effectively been given to variables not selected. Gordon (1980) argues that if no information at all is available about the relevance of different variables, equal weighting would seem appropriate. After a detailed study on a data set has been carried out however, certain variables are likely to be considered of greater importance in discriminating between groups of objects. (De Sarbo *et al.*, 1984, describe a method which in addition to providing a clustering of a set of individuals also renders weights indicating the variable's relative importance to the clustering). Perhaps, as pointed out by Dunn and Everitt (1982), the real problem with *a priori* weighting is not that it is logically invalid, but that it is often very difficult to decide how to weight the variables in practice.

3.3   Similarity measures

A similarity coefficient indicates the strength of the relationship between two objects given the values of a set of p variates common to both. The similarity between two objects i and j, will be some function of their observed values, i.e.

$$s_{ij} = f(\mathbf{x}_i, \mathbf{x}_j) \tag{3.1}$$

where $\mathbf{x}_i' = [x_{i1}, x_{i2}, \ldots, x_{ip}]$ and $\mathbf{x}_j' = [x_{j1}, x_{j2}, \ldots, x_{jp}]$ are the observed variable values for the objects. Many functions have been proposed depending partly on the types of variable concerned (quantitative, categorical, binary, ordinal) and partly on the type of object (see later).

Similarity is usually regarded as a symmetric relationship requiring $s_{ij} = s_{ji}$. (Asymmetric proximity measures are considered in Constantine and Gower, 1978). Most similarity coefficients are non negative and are scaled so as to have an upper limit of unity, although some are of a correlational nature so that

$$-1 \leqslant s_{ij} \leqslant 1.$$

Associated with every similarity measure bounded by zero and unity is a dissimilarity $d_{ij} = 1 - s_{ij}$ which is symmetric and non negative. The degree of similarity between two objects increases with s_{ij} and decreases with increasing d_{ij}. It is natural for an object to have maximal similarity with itself so that $s_{ii} = 1$ and $d_{ii} = 0$.

3.3.1   Similarity measures for binary variables

The simplest and most commonly used similarity coefficients are those for dichotomous variables where each variable has two values only. In some cases

Table 3.1 Counts of binary variables for two individuals

		Individual i		
		1	2	
Individual j	1	a	b	$a+b$
	2	c	d	$c+d$
Total		$a+c$	$b+d$	p

Table 3.2 Similarity coefficients for binary data

$(i)\quad \dfrac{a+d}{p}$ $\qquad (ii)\quad \dfrac{a}{a+b+c}$

$(iii)\quad \dfrac{2a}{2a+b+c}$ $\qquad (iv)\quad \dfrac{2(a+d)}{2(a+d)+b+c}$

$(v)\quad \dfrac{a}{a+2(b+c)}$

these may relate to the presence or absence of some quality, in others simply to two equivalent alternatives, high/low, rough/smooth etc. (As will be seen later this distinction is often an important one). Such data for two individuals, i and j can be arranged in the 2×2 table (Table 3.1). Here, $p = a + b + c + d$. (This 2×2 table as used in clustering applications is primarily a convenient way of arranging the data and should not be confused with the usual 2×2 contingency table).

Many similarity coefficients have been proposed that combine the quantities a, b, c and d, a number of which are listed in Table 3.2. More extensive lists are given in Sneath and Sokal (1973), Anderberg (1973), Clifford and Stephenson (1975), Cormack (1971) and Gower (1985).

The two coefficients most commonly used in practice are the matching coefficient (i) and Jaccard's coefficient (ii). The first is simply the ratio of the total number of variables on which the two individuals match, to the total number of variables; the second is the corresponding ratio when 'negative' matches (d) are ignored. The problem of whether or not to include the latter is of concern only when the variables are of the 'present', 'absent' variety. In such cases it might be unreasonable to consider two individuals as very similar, simply because they both lack a large number of qualities. It is in the application of numerical methods to the classification of living organisms that the difficulty is most pronounced, as the following quotation from Sokal and Sneath (1963) illustrates:

> "The absence of wings, when observed among a group of distantly related organisms (such as a camel, a horse and nematode), would surely be an ab-

surd indication of affinity. Yet a positive character such as the presence of wings (or flying organs defined without qualification as to kind of wing) could mislead equally when considered for a similarly heterogeneous assemblage (for example, bat, heron and dragonfly). Neither can we argue that absence of a character may be due to a multitude of causes and that matched absence in a pair of individuals is therefore not 'true resemblance' for, after all, we know little more about the origins of matched positive characters."

The same authors give a full discussion of similarity coefficients for use with binary data and argue that no hard and fast rule can be made regarding the inclusion or otherwise of negative matches. Each set of data must be considered on its merits by the investigator most familiar with the material involved.

The different similarity coefficients may have widely different values for the same set of data. Suppose, for example, two individuals have the following scores on ten binary variables.

	Variable									
	1	2	3	4	5	6	7	8	9	10
Individual 1	1	0	0	0	1	1	0	0	1	0
Individual 2	0	0	0	0	1	0	0	1	1	0

The corresponding 2×2 table is

		Individual 1		
		1	0	
	1	2	1	3
Individual 2				
	0	2	5	7
		4	6	10

Here the values taken by the various coefficients given in Table 3.2 are (i) 0.70, (ii) 0.40, (iii) 0.57, (iv) 0.82, (v) 0.25. That the different coefficients take different values for the same pair of individuals would be relatively unimportant if the coefficients were *jointly monotonic*, in the sense that, if all the values for different pairs of individuals on one coefficient were ordered so that they formed a monotonic series (that is a series which either increases or decreases throughout its length), the corresponding values for other coefficients were similarly ordered. That this is not necessarily the case is most easily demonstrated by introducing data on the ten binary variables considered previously, for a further individual.

	Variable									
	1	2	3	4	5	6	7	8	9	10
Individual 3	0	0	0	0	0	0	0	1	0	0

Values for the first two coefficients in Table 3.2 for the three pairs of individuals are

Matching coefficient	Jaccard's coefficient
$s_{12} = 0.70$	$s_{12} = 0.40$
$s_{13} = 0.50$	$s_{13} = 0.00$
$s_{23} = 0.80$	$s_{23} = 0.33$

The coefficients are not jointly monotonic.

Categorical data where the variables have more than two levels—eye colour for example—could be dealt with in a similar way to binary data with each level of a variable being regarded as a single binary variable. This is not an attractive approach, however, simply because of the large number of 'negative' matches which will inevitably be involved. A superior method is to allocate a score s_{ijk} of zero or one, to each variable k, depending on whether the two individuals i and j are the same on the variable. The scores for all variables are then simply averaged to give the required similarity coefficient.

$$s_{ij} = \frac{\sum_{k=1}^{p} s_{ijk}}{p} \tag{3.2}$$

3.3.2 Similarity measures for quantitative variables

Quantitative variables could be dealt with by simply converting them into binary variables and using the coefficients described in the previous section. Length, for example, might be transformed into 'below 10 cm' and '10 cm and above'. Such an approach obviously entails a loss of information and it is perhaps preferable to consider similarity measures which can be applied directly. One such measure which has been widely used is equivalent in form to Pearson's product moment correlation coefficient; its use in the context of clustering is however more contentious than its non-controversial role in assessing the linear relationship between pairs of variables. When used as a measure of similarity for two individuals, its calculation involves averaging over the values of different quantitative variables to produce an 'average variable value' for each individual, a procedure dismissed by Jardine and Sibson (1971) as 'absurd.'

It has often been suggested that the correlation coefficient is a useful measure of similarity in those situations where absolute 'size' alone is seen as less important than 'shape'. In classifying animals and plants for example, the absolute sizes of the organisms or their parts are often less important than their shapes. In such cases the investigator requires a similarity coefficient which takes the value unity whenever the set of variable values describing two individuals are parallel, irrespective of their possibly differing levels. The following sets of scores illustrate this distinction more clearly.

Individual 1	10	5	15	3	20
Individual 2	15	10	20	8	25
Individual 3	30	25	35	23	40

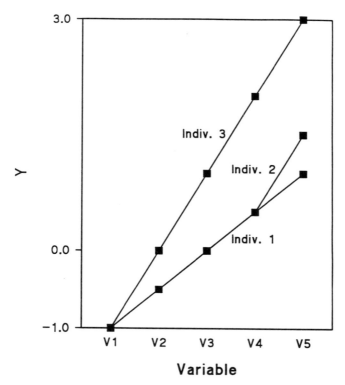

Figure 3.2   Problem with using the correlation coefficient as a similarity coefficient.

The correlation coefficient meets the requirement—it takes the value unity for each pair of individuals in this example. Unfortunately, as pointed out by Fleiss and Zubin (1969) the converse is not true, two individuals may have a correlation of unity even when their profiles of measurements are not parallel. All that is required for perfect correlation is that one set of scores be linearly related to the second set. The three sets of scores illustrated in Figure 3.2 serve to emphasise this point. The values for individual 3 are twice those of individual 1 plus 1. The values for individual 2 are the same as those for individual 1 except on variable 5. The correlation index for individuals 1 and 3 is 1 and for 1 and 2, 0.986. Individuals 1 and 3 are deemed more similar than individuals 1 and 2, a conclusion unlikely to be sensible in any classification exercise.

Some authors have criticised the correlation index for its failure to work in practice. Wishart (1971), for example, found that it was unable to separate correctly the four relatively distinct groups present in a data set constructed from a bivariate normal population. On the other hand Strauss *et al.* (1973) use the correlation coefficient to measure similarity in a study of various clustering techniques applied to psychiatric data, with some success. These authors find that the use of this measure leads to the complete recovery of the five groups in an artificially generated set of data of one hundred individuals measured on forty-eight variables; in contrast the use of the Euclidean distance measure (see Section 3.4), fails to recover the groups.

An interesting discussion of some further aspects of the measurement of similarity is given in Skinner (1978) and a number of other similarity coefficients for quantitative variables are described in Clifford and Stephenson (1975). Other measures could be obtained by simple transformations of the dissimilarity and distance measures to be described in Section 3.4.

3.3.3 Similarity measures for variables of mixed type

The problem of data sets containing a variety of variable types was mentioned in Section 3.2. A similarity coefficient suggested by Gower (1971) is particularly useful for such data sets. It is defined as

$$s_{ij} = \sum_{k=1}^{p} w_{ijk}s_{ijk} \Big/ \sum_{k=1}^{p} w_{ijk} \qquad (3.3)$$

In this formula, s_{ijk} is the similarity between the ith and jth individuals as measured by the kth variable and w_{ijk} is typically 1 or 0 depending on whether or not the comparison is considered valid for the kth variable. Weights of zero are assigned when variable k is unknown for one or both individuals, or to binary variables where it is required to exclude negative matches. For categorical data the component similarities, s_{ijk} take the value one when the two individuals have the same value and zero otherwise. For quantitative variables they are given by

$$s_{ijk} = 1 - |x_{ik} - x_{jk}|/R_k \qquad (3.4)$$

where x_{ik} and x_{jk} are the two individuals' values for variable k, and R_k is the range of variable k, usually in the set of individuals to be clustered.

To illustrate the use of Gower's coefficient consider the following data for five psychiatrically ill patients.

	Weight (pounds)	Anxiety level	Depression present?	Hallucination present?	Age group
Patient 1	120	Mild	No	No	Young
Patient 2	150	Moderate	Yes	No	Middle
Patient 3	110	Severe	Yes	Yes	Old
Patient 4	145	Mild	No	Yes	Old
Patient 5	120	Mild	No	Yes	Young

In this case let us suppose that the investigator wishes to exclude negative matches on depression and hallucinations from the calculation of between patient similarity. The values of Gower's similarity measure for each pair of patients can be calculated from (3.3). For example

$$s_{12} = \frac{1 \times (1 - \frac{30}{40}) + 1 \times 0 + 1 \times 0 + 0 \times 1 + 1 \times 0}{1 + 1 + 1 + 0 + 1} = 0.0625 \qquad (3.5)$$

Table 3.3   Dissimilarity measures

(i) $\sum_{k=1}^{p}(x_{ik} - x_{jk})^2$ Euclidean distance

(ii) $\sum_{k=1}^{p}|x_{ik} - x_{jk}|$ City block

(iii) $\sum_{k=1}^{p}|x_{ik} - x_{jk}|/(x_{ik} + x_{jk})$ 'Canberra' metric

(iv) $\dfrac{\sum_{k=1}^{p}x_{ik}x_{jk}}{\left[\sum_{k=1}^{p}x_{ik}^2 \sum_{k=1}^{p}x_{ik}^2\right]^{\frac{1}{2}}}$ Angular separation

---

The values for all pairs of patients can be conveniently displayed in the following *similarity matrix*, **S**.

$$
\begin{array}{c}
\quad\quad 1 \quad\quad 2 \quad\quad 3 \quad\quad 4 \quad\quad 5 \\
\mathbf{S} = \begin{array}{c} 1 \\ 2 \\ 3 \\ 4 \\ 5 \end{array}
\left(
\begin{array}{ccccc}
1.000 & & & & \\
0.062 & 1.000 & & & \\
0.150 & 0.200 & 1.000 & & \\
0.344 & 0.175 & 0.425 & 1.000 & \\
0.750 & 0.005 & 0.350 & 0.475 & 1.000
\end{array}
\right)
\end{array}
$$

3.4   Dissimilarity and distance measures

Dissimilarity measures have already been introduced as the complement of similarity measures. Some dissimilarity coefficients have the *metric* property that

$$d_{ij} + d_{ik} \geqslant d_{jk} \qquad\qquad (3.6)$$

for all *i*, *j* and *k*, in which case they are generally known as *distance* measures. A number of possible dissimilarity measures are listed in Table 3.3. A more extensive list is given in Gower (1985).

   Perhaps the most commonly used distance measure and the most familiar is Euclidean. Used on the raw data however, it may be very unsatisfactory since its value is largely dependent on the particular scales chosen for the variables. Suppose, for example three children are each measured on two variables, weight (in pounds) and height (in feet) with the following results

	Weight	Height
Child 1	60	3.0
Child 2	65	3.5
Child 3	63	4.0

The Euclidean distances are $d_{12} = 5.02$, $d_{13} = 3.16$ and $d_{23} = 2.06$. If however, height had been measured in inches, the Euclidean distances become $d_{12} = 7.81$, $d_{13} = 12.37$ and $d_{23} = 6.32$. Child 1 is now deemed to be closer to child 2 than child 3. So even when all the variables are uniquely determined except for scale changes, Euclidean distance will not even preserve distant rankings. Because of this variables are generally standardized before calculating Euclidean distance, by taking $z_{ik} = x_{ik}/s_k$ where s_k is the standard deviation of the kth variable. Although this is not without difficulties (see Section 3.1), the Euclidean distances thus calculated are generally far more satisfactory than those calculated from the raw data.

The calculation of Euclidean distance as given in Table 3.3 assumes that the variable values are uncorrelated with one another. In most applications this assumption will not be justified and Euclidean distance may then be a poor measure to use. An alternative is to use *Mahalanobis* D^2, which for two individuals i and j with vectors of measurements $\mathbf{x}_i$ and $\mathbf{x}_j$, is given by

$$d_{ij} = (\mathbf{x}_i - \mathbf{x}_j)'\mathbf{S}^{-1}(\mathbf{x}_i - \mathbf{x}_j) \tag{3.7}$$

The matrix $\mathbf{S}$ in (3.7) is usually taken to be the pooled within groups covariance matrix. In classification investigations this will, *a priori*, be unknown, and the overall covariance matrix will have to be substituted. This is not particularly satisfactory for the reasons discussed earlier with respect to standardization (see Section 3.2).

For binary data, Gower (1966) has shown that

$$d_{ij} = \sqrt{2(1 - s_{ij})} \tag{3.8}$$

can function as Euclidean distance provided the matrix of similarity coefficients is positive semi-definite. Both the simple matching coefficient and Jaccard's coefficient meet this requirement.

Although Euclidean distance is the most widely used in a clustering context, other distance measures have been employed. The city block metric, for example, has been used in anthropology by Czekanowski (1909, 1932) and by Haltenorth (1937) in a study of eight species of the large cats. Carmichael and Sneath (1969) also prefer this measure over Euclidean distance in their TAXMAP clustering procedure (see Chapter 7). Their rationale is that when, for example, two individuals are specified by two variables whose scale units are of equal value, they should have the same distance whether (a) they are two units apart on each variable or (b) they are one unit apart on one variable and three units apart on the other. The use of the city block metric meets this requirement. A situation in which the city block metric is the natural distance measure is identified by Dunn and Everitt (1982) and involves the comparison of amino acids in homologous proteins.

3.5 Between group similarity and distance measures

In the previous sections methods for the measurement of inter-individual similarity and distance have been described. In clustering applications it is also frequently necessary to be able to define such measures *between groups*. Problems which arise in finding suitable definitions include

(i)   *The choice of a summary statistic for each variable to describe the group.* Sensible choices might be proportions for qualitative variables and means for quantitative variables.

(ii)   *Measurement of within-group variation.*

(iii) *Construction of a measure of similarity or distance* based on (i) and perhaps making allowance for (ii). Making allowance for within-group variation might be particularly tricky if this is not constant from one group to another, and there is no reason to believe that it should be.

3.5.1   Quantitative variables

A distance measure that geneticists have used when describing groups or populations in terms of gene frequencies is the so called *genetic distance*, d_{AB} defined as follows:

$$d_{AB} = (1 - cos\theta)^{\frac{1}{2}} \qquad (3.9)$$

where

$$cos\,\theta = \sum_i (p_{iA}p_{iB})^{\frac{1}{2}} \qquad (3.10)$$

The terms p_{iA} and p_{iB} are the gene frequencies for the ith allele at a given locus in the two populations. The angular transformation for the proportion has a variance-stabilizing role. When several genetic loci are considered the d_{AB} values are added together. This is equivalent to the construction of a city block metric.

   This approach to measuring distances between groups can be generalized to qualitative variables merely by replacing the word 'locus' in the above definition by 'variable' and 'allele' by 'variable category'. If there are p qualitative variables there will be p, d_{AB} values to add together. As an example, consider the set of proportions for two hypothetical populations of red campion shown in Table 3.4.

   For corolla colour:

$$d_{AB} = [1 - (0.95 \times 0.80)^{\frac{1}{2}} - (0.05 \times 0.20)^{\frac{1}{2}}]^{\frac{1}{2}}$$
$$= 0.17$$

For coronal scale colour:

$$d_{AB} = [1 - (0.85 \times 0.75)^{\frac{1}{2}} - (0.01 \times 0.15)^{\frac{1}{2}} - (0.14 \times 0.10)^{\frac{1}{2}}]^{\frac{1}{2}}$$
$$= 0.21$$

For red calyx pigment:

$$d_{AB} = [1 - (0.80 \times 0.60)^{\frac{1}{2}} - (0.20 \times 0.40)^{\frac{1}{2}}]^{\frac{1}{2}}$$
$$= 0.16$$

Table 3.4 Characteristics of two hypothetical populations of red campion

Character state	Proportion	
	Population A	Population B
Corolla colour		
pink	0·95	0·80
white	0·05	0·20
Coronal scale colour		
as petals, pink	0·85	0·75
not as petals, pink	0·01	0·15
not as petals, white	0·14	0·10
Red calyx pigment		
present	0·80	0·60
absent	0·20	0·40

The total distance between the two groups is taken as the sum of the distances for each variable to give

$$d_{AB} = 0.17 + 0.21 + 0.16 = 0.54$$

3.5.2 Quantitative variables

One obvious method for constructing between group distance measures is to simply substitute *group means* for the p variables in the formulae for inter-individual measures such as Euclidean distance or city block distance. If, for example, group A, has mean vector $\bar{\mathbf{x}}'_A = [\bar{x}_{A1}, \bar{x}_{A2}, \cdots, \bar{x}_{Ap}]$ and group B mean vector $\bar{\mathbf{x}}'_B = [\bar{x}_{B1}, \bar{x}_{B2}, \cdots, \bar{x}_{Bp}]$, then one measure of the distance between the two groups would be

$$d_{AB} = \sqrt{\sum_{i=1}^{p} (\bar{x}_{Ai} - \bar{x}_{Bi})^2} \tag{3.11}$$

More appropriate however might be measures which incorporate in one way or another, knowledge of within-group variation. One possibility is Mahalanobis D^2 adapted from the form given in (3.7) to

$$D^2 = (\bar{\mathbf{x}}_A - \bar{\mathbf{x}}_B)' \mathbf{W}^{-1} (\bar{\mathbf{x}}_A - \bar{\mathbf{x}}_B) \tag{3.12}$$

Here $\mathbf{W}$ is a $p \times p$ matrix of pooled within-group dispersions for the two groups. When correlations between variables are slight, D^2 will be similar to the squared Euclidean distance calculated on the standardized data.

Use of D^2 implies that the investigator is willing to assume that the variable dispersions are at least approximately the same in the two groups. When this is not so, D^2 is an inappropriate between-group measure, and in such cases an alternative is the *information radius* suggested by Jardine and Sibson (1971).

This is given by

$$R_{AB} = \log\left[\frac{\det(\frac{1}{2}(\mathbf{W}_B))}{\sqrt{\det\mathbf{W}_A)(\det\mathbf{W}_B)}} + \frac{1}{2}\log_2(1 + \frac{1}{4}D_{AB}^2)\right] \qquad (3.13)$$

where

$$D_{AB}^2 = (\bar{\mathbf{x}}_A - \bar{\mathbf{x}}_B)'\left[\frac{1}{2}(\mathbf{W}_A + \mathbf{W}_B)\right]^{-1}(\bar{\mathbf{x}}_A - \bar{\mathbf{x}}_B) \qquad (3.14)$$

When $\mathbf{W}_A = \mathbf{W}_B$, the first term in the expression for R_{AB} vanishes and D_{AB}^2 is exactly Mahalanobis D^2.

A number of other possibilities for between-group measures are available which are not based simply on substituting group means in inter-individual measures. For example, the distance between two groups could be defined as the distance between their closest members, one from each group. This is sometimes known as *nearest-neighbour distance* and is the basis of the clustering technique known as *single linkage*, to be described in the next chapter. A further possibility which is the opposite of nearest-neighbour distance is to define distance between groups as that between the most remote pair of individuals, one in each group. This is known as *furthest neighbour distance* and is associated with the *complete linkage* cluster method, also to be described in the next chapter. Another inter-group measure may be obtained by taking the average of all the inter-individual measures of those pairs of individuals when the members of the pairs are in different groups. Such a measure is used in *group average clustering*. Lance and Williams (1967) point out that the concept of an average for similarity coefficients is not always acceptable and suggest that a more satisfactory inter-group similarity measure can be obtained from

$$s_{AB} = \cos\left[\frac{1}{n_1 n_2}\sum_{\substack{i \in A \\ j \in B}}\cos^{-1} s_{ij}\right] \qquad (3.15)$$

where s_{AB} is the similarity of groups A and B, n_1 and n_2 are the numbers of individuals in these groups, and s_{ij} represents a single inter-individual measure. This suggestion does not appear to have been used in any clustering application.

More will be said about inter-group distance and similarity measures in the next chapter.

3.6   The graphical display of distance matrices

It is often helpful if the information about inter-individual similarities or distances in a similarity or distance matrix can be presented graphically, usually by deriving coordinate values for the individuals, which may then be plotted. Such plots might then be examined for evidence of distinct clusters of points, or they might be used in association with the results of some clustering technique (see the next chapter). The most common approach to finding a coordinate representation of a set of distances or similarities is by using one or other of a collection of techniques known as *multidimensional scaling*. Such methods are described in detail in Everitt and Dunn (1991). In the next section a relatively brief account is given along with a number of examples.

3.6.1 Multidimensional scaling

A geometrical or spatial representation of the observed proximity matrix consists of a set of points $x_1, x_2, \ldots, x_n$ in p dimensions, each point representing one of the individuals or objects under investigation, and a measure of distance between pairs of points. The object of multidimensional scaling is to determine both the dimensionality needed to represent the information in the proximity matrix adequately, and the position of the points, so that there is, in some sense, maximum correspondence between the observed proximities and the interpoint distances. In general terms this simply means that the larger the observed distance between two individuals the further apart should be the points representing them.

The required coordinates are found by minimizing some function which measures the discrepancy between the observed proximities and the fitted distances. Many such functions have been suggested, a number of which are described in Everitt and Dunn (1991). In all cases the results will consist of a set of coordinate values for each individual. The hope is that the first few of these will provide an adequate representation of the observed proximities. As a first example consider the airline distances between ten cities in the USA given in Table 3.5. The two-dimensional coordinates obtained from applying classical (*metric*) multidimensional scaling (see Everitt and Dunn, 1991), are shown in Table 3.6; and the resulting two-dimensional plot in Figure 3.3. Clearly this diagram provides a very adequate description of the inter-city distances.

A more interesting application of multidimensional scaling in a clustering context is an example given by Prentice (1979, 1980), in an investigation of the genus *Silene*. Jardine and Sibson's information radius (see previous section) was calculated for all pairs in 97 European populations of the genus, and *non-metric* multidimensional scaling applied. (See Kruskal and Wish, 1978). The resulting two-dimensional diagram is shown in Figure 3.4. Two distinct clusters are seen corresponding to the two species *Silene alba* and *Silene dioica*. The scatter plot also shows that variation in the former species is far greater than in the latter.

3.7 Summary

Questions which arise in the measurement of similarity and distance are numerous and have been discussed only relatively briefly in this chapter. (The issues are more fully considered in Williams and Dale, 1965, Morrison, 1967, Sneath and Sokal, 1973, Anderberg, 1973, and Clifford and Stephenson, 1975). The difficulties for clustering techniques are however clear—which similarity or distance measure should be used since different measures may lead to different results? Unfortunately and despite a number of comparative studies (see Cheetham and Hazel, 1969, Boyce, 1969, and Williams, Lambert and Lance,

Figure 3.3   Multidimensional scaling solution on airline distances.

Table 3.5   Airline distances between 10 US cities (Kruskal and Wish, 1978, courtesy of Sage Publications)

	Atla	Chic	Denv	Hous	LA	Mia	NY	SF	Seat	Wash
Atlanta	—									
Chicago	587	—								
Denver	1212	920	—							
Houston	701	940	879	—						
Los Angeles	1936	1745	831	1374	—					
Miami	604	1188	1726	968	2339	—				
New York	748	713	1631	1420	2451	1092	—			
San Francisco	2139	1858	949	1645	347	2594	2571	—		
Seattle	2182	1737	1021	1891	959	2734	2408	678	—	
Washington DC	543	597	1494	1220	2300	923	205	2442	2329	—

1966), the question cannot be answered in any absolute sense and the choice of measure will have to be guided largely by the type of variables being used and the intuition of the investigator. One recommendation which appears sensible however, is that of Sneath and Sokal (1973) who suggest that the *simplest* coefficient applicable to a data set be chosen, since this is likely to ease the possibly difficult task of interpretation of final results.

Table 3.6 Eigenvalues and eigenvectors arising from classical multidimensional scaling applied to distances in Table 3.5

	Eigenvalues		First two eigenvectors	
			1	2
1	9582144·3	Atlanta	718·7	− 143·0
2	1686820·1	Chicago	382·0	340·8
3	8157·0	Denver	− 481·6	25·3
4	1432·9	Houston	161·5	− 527·8
5	507·7	Los Angeles	− 1203·7	− 390·1
6	25·1	Miami	1133·5	− 581·9
7	0·0	New York	1072·2	519·0
8	− 897·7	San Francisco	− 1420·6	− 112·6
9	− 5467·6	Seattle	− 1341·7	579·7
10	− 35478·9	Washington DC	979·6	335·5

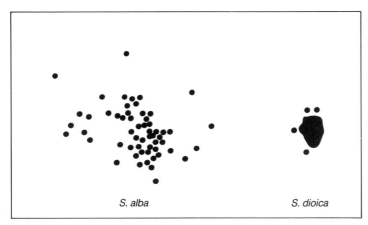

Figure 3.4 Non-metric multidimensional scaling of the genus *Silene*. (Taken with permission from Prentice, 1979.)

4

Hierarchical Clustering Techniques

4.1 Introduction

In a *hierarchic* classification the data are not partitioned into a particular number of classes or clusters at a single step. Instead the classification consists of a series of partitions which may run from a single cluster containing all individuals, to n clusters each containing a single individual. Hierarchical clustering techniques may be subdivided into *agglomerative* methods which proceed by a series of successive fusions of the n individuals into groups, and *divisive* methods, which separate the n individuals successively into finer groupings. Both types of hierarchical clustering can be viewed as attempting to find the most efficient step, in some defined sense (see later), at each stage in the progressive subdivision or synthesis of the data. With such methods, divisions or fusions once made are irrevocable, so that when an agglomerative algorithm has joined two individuals they cannot subsequently be separated, and when a divisive algorithm has made a split it cannot be reunited. As Kaufman and Rousseeuw (1990) colourfully comment 'A hierarchical method suffers from the defect that it can never repair what was done in previous steps.' A useful illustration of the problem is provided by Hawkins, Muller and ten Krooden (1982). Since all agglomerative hierarchical techniques ultimately reduce the data to a single cluster containing all the individuals, and the divisive techniques will finally split the entire set of data into n groups each containing a single individual, the investigator wishing to have a solution with an 'optimal' number of clusters, will need to decide on a particular stage to stop. The problem of deciding on the correct number of clusters is discussed fully in Section 4.6.

Hierarchic classifications may be represented by a two-dimensional diagram known as a *dendrogram* which illustrates the fusions or divisions made at each successive stage of the analysis. An example of such a diagram is given in Figure 4.1; others will be given later. The structure in Figure 4.1 resembles

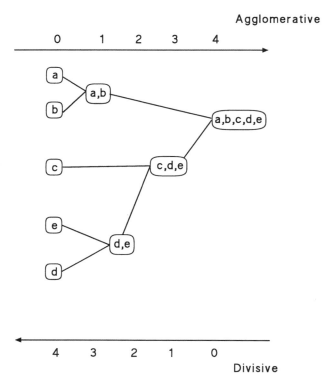

Figure 4.1 Example of a dendrogram. (Reproduced with permission from Kaufman and Rosseeuw, 1990.)

an *evolutionary tree* (see Figure 4.2) and it is in biological applications that hierarchical classifications are perhaps most relevant (although, as will be seen later, hierarchical clustering methods have been applied in many other areas). According to Rohlf (1970), a biologist, 'all things being equal' aims for a system of nested clusters. Hawkins *et al.* (1982) however, issue the following caveat: 'users should be very wary of using hierarchic methods if they are not clearly necessary'.

4.2   Agglomerative methods

An agglomerative hierarchical clustering procedure produces a series of partitions of the data, P_n, P_{n-1}, ..., P_1. The first, P_n, consists of n single-member 'clusters,' the last P_1, consists of a single group containing all n individuals. The basic operation of all such methods is similar, and is outlined in Figure 4.3.

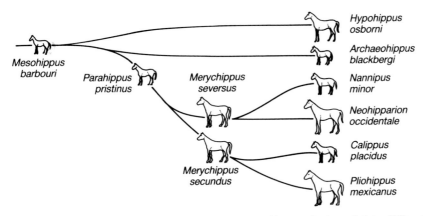

Figure 4.2 An evolutionary tree. (Reproduced with permission of John Wiley & Sons from Kaufman and Rosseeuw, *Finding Groups in Data*, 1990.)

START: Clusters C_1, C_2, ..., C_n each containing a single individual.

1. Find nearest pair of distinct clusters, say C_i and C_j, merge C_i and C_j, delete C_j and decrement number of clusters by one.

If number of clusters equal one then stop, else return to 1.

Figure 4.3 Basic operation of hierarchical clustering prodecures.

At each particular stage the methods fuse individuals or groups of individuals which are closest (or most similar). Differences between methods arise because of the different ways of defining distance (or similarity) between an individual and a group containing several individuals, or between two groups of individuals. Several agglomerative hierarchical techniques will now be described in detail and for convenience this description will be in terms of distance measures.

4.2.1 Single linkage clustering

One of the simplest agglomerative hierarchical clustering methods is *single linkage* also often known as the *nearest neighbour* technique. It was first described by Florek *et al.* (1951) and later by Sneath (1957) and Johnson (1967). The defining feature of the method is that distance between groups is defined as that of the closest pair of individuals, where only pairs consisting of one individual from each group are considered. This measure of inter-group distance is illustrated in Figure 4.4.

As an example of the operation of agglomerative hierarchical techniques in general and single linkage in particular, the method will be applied to the

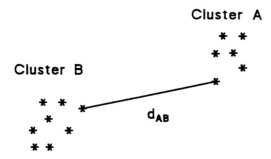

Figure 4.4 Single linkage distance.

following distance matrix.

$$
\mathbf{D}_1 = \begin{array}{c} \\ 1 \\ 2 \\ 3 \\ 4 \\ 5 \end{array}
\begin{array}{ccccc}
1 & 2 & 3 & 4 & 5 \\
\left(\begin{array}{ccccc}
0.0 & & & & \\
2.0 & 0.0 & & & \\
6.0 & 5.0 & 0.0 & & \\
10.0 & 9.0 & 4.0 & 0.0 & \\
9.0 & 8.0 & 5.0 & 3.0 & 0.0
\end{array} \right)
\end{array}
$$

The smallest entry in the matrix is that for individuals 1 and 2, consequently these are joined to form a two-member cluster. Distances between this cluster and the other three individuals are obtained as

$$
\begin{aligned}
d_{(12)3} &= \min[d_{13}, d_{23}] = d_{23} = 5.0 \\
d_{(12)4} &= \min[d_{14}, d_{24}] = d_{24} = 9.0 \\
d_{(12)5} &= \min[d_{15}, d_{25}] = d_{25} = 8.0
\end{aligned}
\tag{4.1}
$$

A new matrix may now be constructed whose entries are inter-individual distances *and* cluster-individual values.

$$
\mathbf{D}_2 = \begin{array}{c} \\ (12) \\ 3 \\ 4 \\ 5 \end{array}
\begin{array}{cccc}
(12) & 3 & 4 & 5 \\
\left(\begin{array}{cccc}
0.0 & & & \\
5.0 & 0.0 & & \\
9.0 & 4.0 & 0.0 & \\
8.0 & 5.0 & 3.0 & 0.0
\end{array} \right)
\end{array}
$$

The smallest entry in $\mathbf{D}_2$ is that for individuals 4 and 5, so these now form a second two-member cluster, and a new set of distances found

$$
\begin{aligned}
d_{(12)3} &= 5.0 \text{ as before} \\
d_{(12)(45)} &= \min[d_{14}, d_{15}, d_{24}, d_{25}] = d_{25} = 8.0 \\
d_{(45)3} &= \min[d_{34}, d_{35}] = d_{34} = 4.0
\end{aligned}
\tag{4.2}
$$

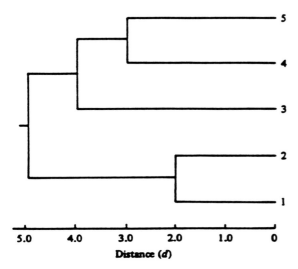

Figure 4.5 Single linkage dendrogram.

These may be arranged in a matrix $\mathbf{D}_3$:

$$\mathbf{D}_3 = \begin{array}{c} \\ (12) \\ 3 \\ (45) \end{array} \begin{array}{ccc} (12) & 3 & (45) \\ \left(\begin{array}{ccc} 0.0 & & \\ 5.0 & 0.0 & \\ 8.0 & 4.0 & 0.0 \end{array} \right) \end{array}$$

The smallest entry is now $d_{(45)3}$ and so individual 3 is added to the cluster containing individuals 4 and 5. Finally the groups containing individuals 1, 2 and 3, 4, 5 are combined into a single cluster. The partitions produced at each stage are as follows:

Stage	Groups
P_5	[1], [2], [3], [4], [5]
P_4	[1 2], [3], [4], [5]
P_3	[1 2], [3], [4 5]
P_2	[1 2], [3 4 5]
P_1	[1 2 3 4 5]

The corresponding dendrogram is shown in Figure 4.5.

An important point to note about the results is that the clusterings proceed hierarchically, each being obtained by the merger of clusters from the previous level. So, for example, at the fourth stage clusters (1, 2, 4) and (3, 5) could not have been formed since neither are obtainable by merger of clusters present at the preceding stage.

The procedure outlined above is useful for illustrating the operation of single linkage clustering on a small data set, but is more complicated than it need

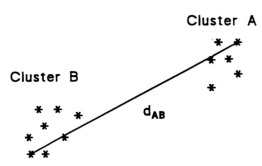

Figure 4.6   Complete linkage distance.

be. The single linkage dendrogram can be found very simply by examining the entries in $\mathbf{D}_1$ as follows.

(i)   Look for the smallest distance—d_{12}—and cross that off in the table.
(ii)   Look for the next smallest distance—d_{45}—and again cross it off.
(iii)   Look for the next smallest distance—d_{43}—linking in individual 3 with individuals 4 and 5, and again cross the corresponding distance off.
(iv)   Only individual 5 is left and this links them all together.

Very efficient methods which allow the single linkage method to be applied to large data sets are described by Sibson (1973) and Rohlf (1973, 1978).

The single linkage method is closely related to certain aspects of graph theory. A graph is a set of nodes and of relations between pairs of nodes indicated by joining the nodes. A set of observations and their dissimilarities may be represented in a graph as nodes and edges respectively. A spanning tree of a graph is a set of edges which provides a unique path between every pair of nodes. A minimal spanning tree (see Chapter 2) is the shortest of all spanning trees. Zahn (1971) gives a family of graph theoretical clustering algorithms based on the minimal spanning tree.

4.2.2   Complete linkage clustering

The *complete linkage* or *furthest neighbour* clustering method is the opposite of single linkage in the sense that distance between groups is now defined as that of the most distant pair of individuals, one from each group. The measure is illustrated in Figure 4.6.

Using this method on the matrix, $\mathbf{D}_1$ of the previous section, the first stage is again the merger of individuals 1 and 2. The distances between this group and the three remaining individuals now becomes

$$d_{(12)3} = \max[d_{13}, d_{23}] = d_{13} = 6.0$$
$$d_{(12)4} = \max[d_{14}, d_{24}] = d_{14} = 10.0 \qquad (4.3)$$
$$d_{(12)5} = \max[d_{15}, d_{25}] = d_{15} = 9.0$$

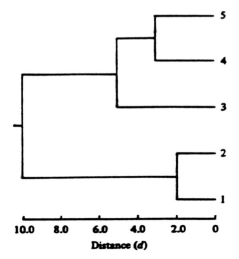

Figure 4.7 Complete linkage dendrogram.

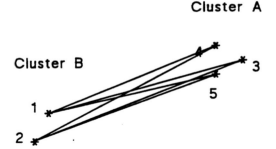

$$d_{AB}=(d_{13}+d_{14}+d_{15}+d_{23}+d_{24}+d_{25})/6$$

Figure 4.8 Group average distance.

The dendrogram obtained from the application of complete linkage to D_1 is shown in Figure 4.7.

4.2.3 Group-average clustering

Here the distance between two clusters is defined as the average of the distances between all pairs of individuals that are made up of one individual from each group. Such a measure is illustrated in Figure 4.8.

Applying the method to the matrix D_1 given in section 4.2.1, the first stage, as with single and complete linkage is the formation of a cluster containing individuals 1 and 2. A new set of distances is found from

$$d_{(12)3} = \frac{1}{2}(d_{13} + d_{23}) = 5.5$$

$$d_{(12)4} = \frac{1}{2}(d_{14} + d_{24}) = 9.5 \qquad (4.4)$$

$$d_{(12)5} = \frac{1}{2}(d_{15} + d_{25}) = 8.5$$

Arranging these in a matrix $\mathbf{D}_2$ gives

$$\mathbf{D}_2 = \begin{matrix} & (12) & 3 & 4 & 5 \\ (12) & \begin{pmatrix} 0.0 & & & \\ 3 & 5.5 & 0.0 & & \\ 4 & 9.5 & 4.0 & 0.0 & \\ 5 & 8.5 & 5.0 & 3.0 & 0.0 \end{pmatrix} \end{matrix}$$

The smallest entry is d_{45} and so a second cluster is formed from individuals 4 and 5. The group average distance between the two, two-member clusters is given by

$$d_{(12)(45)} = \frac{1}{4}(d_{14} + d_{15} + d_{24} + d_{25}) = 9.0 \qquad (4.5)$$

and the procedure would continue as described in previous sections.

The three methods described above operate directly on the proximity matrix and do not need access to the original variable values of the individuals. A method which does require the original data (but see section 4.3) is centroid clustering.

4.2.4   Centroid clustering

With this method, groups once formed are represented by their mean values for each variable, that is, their *mean vector*, and inter-group distance is now defined in terms of distance between two such mean vectors (see Chapter 3). Use of a mean implies strictly, that the variables are on an interval scale; the method is however, often used for other types of variables as will be seen later.

To illustrate how centroid clustering operates it will be applied to the following set of bivariate data.

Individual	Variable 1	Variable 2
1	1.0	1.0
2	1.0	2.0
3	6.0	3.0
4	8.0	2.0
5	8.0	0.0

A plot of the data is shown in Figure 4.9.

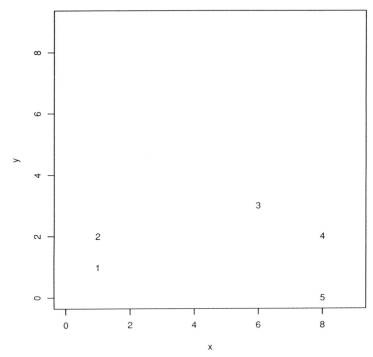

Figure 4.9 Two-dimensional data set.

Suppose Euclidean distance is chosen as the inter-individual measure giving the following distance matrix

$$
\mathbf{D}_1 = \begin{array}{c}
\begin{array}{ccccc} 1 & 2 & 3 & 4 & 5 \end{array} \\
\begin{array}{c} 1 \\ 2 \\ 3 \\ 4 \\ 5 \end{array}
\left(\begin{array}{ccccc}
0.00 & & & & \\
1.00 & 0.00 & & & \\
5.39 & 5.10 & 0.00 & & \\
7.07 & 7.00 & 2.24 & 0.00 & \\
7.07 & 7.28 & 3.61 & 2.00 & 0.00
\end{array}\right)
\end{array}
$$

Examination of $\mathbf{D}_1$ shows that d_{12} is the smallest entry and individuals 1 and 2 are fused to form a group. The mean vector of the group is calculated, (1.0, 1.5), and a new Euclidean distance matrix calculated.

$$
\mathbf{D}_2 = \begin{array}{c}
\begin{array}{cccc} (12) & 3 & 4 & 5 \end{array} \\
\begin{array}{c} (12) \\ 3 \\ 4 \\ 5 \end{array}
\left(\begin{array}{cccc}
0.00 & & & \\
5.22 & 0.00 & & \\
7.02 & 2.24 & 0.00 & \\
7.16 & 3.61 & 2.00 & 0.00
\end{array}\right)
\end{array}
$$

The smallest entry in $\mathbf{D}_2$ is d_{45} and individuals 4 and 5 are therefore fused to

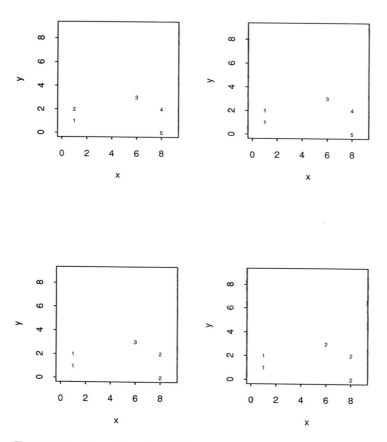

Figure 4.10   Stages in centroid clustering.

form a second group, the mean vector of which is then calculated, giving (8.0, 1.0). A further distance matrix, D_3 is now calculated.

$$D_3 = \begin{matrix} & \begin{matrix} (12) & (45) \end{matrix} \\ \begin{matrix} (12) \\ 3 \\ (45) \end{matrix} & \begin{pmatrix} 0.00 & & \\ 5.22 & 0.00 & \\ 7.02 & 2.83 & 0.00 \end{pmatrix} \end{matrix}$$

In D_3 the smallest entry is $d_{(45)3}$ and so individuals 3, 4 and 5 are merged into a three-member cluster. The final stage consists of the fusion of the two remaining groups into one. The stages are illustrated in Figure 4.10 and the resulting dendrogram appears in Figure 4.11.

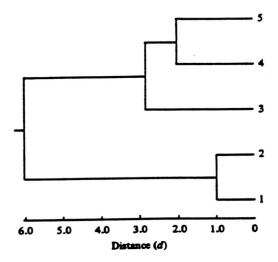

Figure 4.11 Centroid clustering dendrogram.

4.2.5   Median clustering

A disadvantage of the centroid method is that if the sizes of the two groups to be fused are very different then the centroid of the new group will be very close to that of the larger group and may remain within that group. The characteristic properties of the smaller group are then virtually lost. The strategy can be made independent of group size by assuming that the groups to be fused are of equal size, the apparent position of the new group will then always be between the two groups to be fused. Moreover if the centroids of the groups to be fused are represented by (i) and (j), then the distance of the centroid of a third group (h) from the group formed by the fusion of (i) and (j) lies along the median of the triangle defined by $(i), (j)$ and (h). It is for this reason that Gower (1967b), who first suggested the strategy, proposed the name *median*.

Although this method could be made suitable for both similarity and distance measures, Lance and Williams (1967) suggest that it should be regarded as unsuitable for measures such as correlation coefficients, since interpretation in a geometrical sense is no longer possible.

4.2.6   Ward's hierarchical clustering method

Ward (1963) proposed a clustering procedure seeking to form the partitions P_n, $P_{n-1}, \ldots, P_1$ in a manner that minimizes the loss associated with each grouping, and to quantify that loss in a form that is readily interpretable. At each step in the analysis, union of every possible pair of clusters is considered and the two clusters whose fusion results in the minimum increase in 'information loss' are combined. Information loss is defined by Ward in terms of an error sum-of-squares criterion, ESS.

The rationale behind Ward's proposal can be illustrated most simply by considering univariate data. Suppose for example, 10 individuals have scores (2, 6, 5, 6, 2, 2, 2, 0, 0, 0) on some particular variable. The loss of information that would result from treating the ten scores as one group with a mean of 2.5 is represented by ESS given by,

$$\text{ESS} = \sum_{i=1}^{n} (x_i - \bar{x})^2 \tag{4.6}$$

For this example

$$\text{ESS}_{\text{one group}} = (2 - 2.5)^2 + (6 - 2.5)^2 + \ldots + (0 - 2.5)^2 = 50.5 \tag{4.7}$$

Similarly if the 10 individuals are classified according to their scores into four sets,

$$\{0, 0, 0\}, \{2, 2, 2, 2\}, \{5\}, \{6, 6\}$$

the ESS can be evaluated as the sum of four separate error sums of squares

$$\text{ESS}_{\text{four groups}} = \text{ESS}_{\text{group1}} + \text{ESS}_{\text{group2}} + \text{ESS}_{\text{group3}} + \text{ESS}_{\text{group4}} = 0.0 \tag{4.8}$$

Examples of the application of Ward's method are given later.

4.2.7   Other hierarchical techniques

The hierarchical clustering procedures described previously are those that are most widely used in practice. A number of other methods have however been described, several of which merit more investigation and wider use. Saito (1980), for example, suggests a clustering scheme for situations in which a number of individuals rank a number of objects. Statistically homogeneous clusters are produced on the basis of Kendall's coefficient of concordance. Gowda and Krishna (1978) propose a method based on the concepts of *mutual nearest neighbours* and *mutual neighbourhood value* defined for each individual. The latter, for example, is derived as follows.

Let individual k be the mth nearest neighbour of individual j (using some particular metric, usually Euclidean), and individual j be the lth nearest neighbour of individual k. Then the mutual neighbourhood value between individuals j and k is defined as $m + l$.

The examples given in the paper demonstrate that the method can uncover types of cluster structure that would raise considerable difficulties for more conventional methods. Rohlf (1970), Lukasova (1979) and Hansen and Tukey (1992) describe other possible hierarchical methods.

4.3   Lance and Williams' recurrence formula

Lance and Williams (1967) derive a recurrence formula which gives the distance between a group k and a group, (ij) formed by the fusion of groups i and j, as

$$d_{k(ij)} = \alpha_i d_{ki} + \alpha_j d_{kj} + \beta d_{ij} + \gamma |d_{ki} - d_{kj}| \tag{4.9}$$

where d_{ij} is the distance between groups i and j. The inter-group distance measures used by the hierarchical clustering techniques described in the previous section can, by suitable choice of the parameters, $\alpha_i, \alpha_j, \beta$ and γ, be contained within this formula. Single linkage, for example, corresponds to the parameter values $\alpha_i = \alpha_j = \frac{1}{2}, \beta = 0$ and $\gamma = -\frac{1}{2}$. This is easily demonstrated as follows; for the appropriate parameter values (4.9) leads to

$$d_{k(ij)} = \frac{1}{2}d_{ki} + \frac{1}{2}d_{kj} - \frac{1}{2}|d_{ki} - d_{kj}| \tag{4.10}$$

If $d_{ki} > d_{kj}$ then $|d_{ki} - d_{kj}| = d_{ki} - d_{kj}$, and the formula leads to

$$d_{k(ij)} = d_{kj} \tag{4.11}$$

Similarly if $d_{ki} < d_{kj}$ then $|d_{ki} - d_{kj}| = d_{kj} - d_{ki}$, and consequently the recurrence formula gives the required

$$d_{k(ij)} = \min[d_{ki}, d_{kj}] \tag{4.12}$$

For complete linkage the parameter values are $\alpha_i = \alpha_j = \frac{1}{2}, \beta = 0$ and $\gamma = \frac{1}{2}$, and for group average

$$\alpha_i = \frac{n_i}{n_i + n_j}; \alpha_j = \frac{n_j}{n_i + n_j}; \beta = 0; \gamma = 0 \tag{4.13}$$

For these three methods the formulae can be used when d_{ij} represents distances or similarities.

If d_{ij} is Euclidean distance then it can be shown that centroid clustering can be fitted into the scheme with parameter values

$$\alpha_i = \frac{n_i}{n_i + n_j}; \alpha_j = \frac{n_j}{n_i + n_j}; \beta = -\alpha_i\alpha_j; \gamma = 0 \tag{4.14}$$

Wishart (1969b) demonstrates that Ward's method can also be included if d_{ij} represents *squared* Euclidean distance in which case the relevant parameter values are

$$\alpha_i = \frac{n_k + n_i}{n_k + n_i + n_j}; \alpha_j = \frac{n_k + n_j}{n_k + n_i + n_j}; \beta = \frac{-n_k}{n_k + n_i + n_j}; \gamma = 0 \tag{4.15}$$

Lance and Williams (1967) suggest a further hierarchical clustering scheme based on the recurrence relationship above, with parameter values

$$\alpha_i + \alpha_j + \beta = 1; \alpha_i = \alpha_j; \beta < 1; \gamma = 0 \tag{4.16}$$

By allowing β to vary, clustering schemes with various characteristics can be obtained. Lance and Williams suggest that most useful are small negative values of β, and in their examples use $\beta = -0.25$. More recent work by Scheibler and Schneider (1985) and Milligan (1989) suggests that other values of β might be more successful in recovering the underlying cluster structure. The first authors, for example, found that the best recovery was achieved with $\beta = -0.50$.

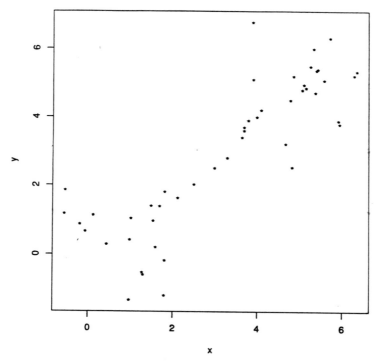

Figure 4.12   Two well separated clusters with intermediate 'noise' points.

4.4   Properties and problems of agglomerative hierarchical clustering techniques

Several hierarchical clustering techniques such as single linkage and the median method have a tendency to cluster together, at relatively low level, individuals linked by a series of intermediates. This property, known generally as *chaining* may cause the methods to fail to resolve relatively distinct clusters when there are a small number of individuals lying between them. (Such points are referred to as *noise* points by Wishart, 1969a). As an example consider the data shown in Figure 4.12. The single linkage dendrogram for the Euclidean distance matrix of these data is given in Figure 4.13, and the two group solution obtained from this dendrogram is displayed in Figure 4.14. Clearly these groups do not correspond to those suggested by Figure 4.12.

Chaining is generally regarded as a defect of single linkage but, as Jardine and Sibson (1968) point out, to call it a defect is rather misleading, chaining is simply a description of what the method does. In some cases it may lead to a more accurate picture of the structure in the data than other methods (see later). A number of methods designed to overcome the chaining difficulty will be described in Chapter 7.

Several of the hierarchical methods described above are biased towards find-

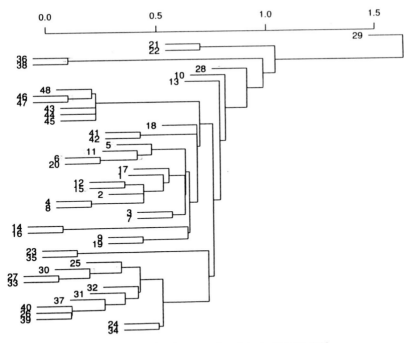

Figure 4.13 Single linkage dendrogram for data in Figure 4.12.

ing 'spherical' clusters even when the data contain clusters of other shapes. To illustrate this problem, single linkage, centroid and Ward's clustering were applied to fifty bivariate observations, produced by sampling twenty five observations from each of two bivariate normal populations with mean vectors $\mu_1 = [0.0, 0.0]$ and $\mu_2 = [4.0, 4.0]$ and common covariance matrix

$$\Sigma = \begin{pmatrix} 16.0 & 1.5 \\ 1.5 & 0.25 \end{pmatrix}$$

The data contain two well separated 'elliptical' groups (see Chapter 5). The means of the two main groups found by each method, ignoring the two or three single member groups given by both single linkage and centroid clustering, are as follows:

(i) Single linkage

Mean	Variable 1	Variable 2	No. of observations
Group 1	−0.5	0.3	23
Group 2	3.5	3.8	24

(ii) Centroid clustering

Mean	Variable 1	Variable 2	No. of observations
Group 1	−1.4	0.15	17
Group 2	3.8	3.0	30

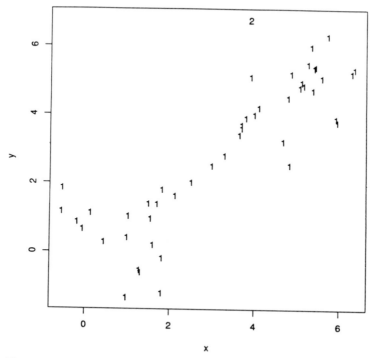

Figure 4.14 Single linkage two group solution for data in Figure 4.13.

(iii) Ward's method

Mean	Variable 1	Variable 2	No. of observations
Group 1	−2.8	0.2	13
Group 2	3.8	2.6	37

Here only single linkage has recovered the real structure in the data. Both centroid clustering and Ward's method have 'imposed' a spherical solution. (Compare the results given in Chapter 5).

Jardine and Sibson (1971) object to many of the agglomerative hierarchical methods described in Section 4.2 on mathematical grounds. Briefly what these authors show is that a cluster method which transforms a dissimilarity coefficient into a hierarchic dendrogram may be regarded as a method whereby the *ultrametric* inequality—$d(x, y) \leqslant \max[d(x, z), d(z, y)]$—is imposed on the coefficient which orginally may have satisfied only the weaker triangular inequality (see Chapter 3). They then specify certain simple conditions that any such transformation should satisfy such as continuity, minimum distortion etc., and then demonstrate that only single linkage clustering satisfies all such conditions. Consequently Jardine and Sibson recommend the method as that of greatest mathematical appeal. Such a recommendation has not been greeted with universal acceptance primarily because, in many applications, single linkage is often the *least* successful method in producing useful cluster solutions. So, for example, Williams *et al.* (1971) question the need for cluster methods to satisfy all of Jardine and Sibson's proposed criteria, and adopt a more pragmatic approach to clustering, insisting that in practice single linkage fails to produce useful solutions. Again, Gower (1975) feels that Jardine and Sibson's rejection of all but single linkage clustering is too extreme and questions whether their criteria are not too stringent. In a later paper, Gower (1988), this point is made explicit:

> "That Sibson showed that only single-linkage clustering could simultaneously satisfy all his axioms certainly gave a nice characterisation of the method, but to my mind, indicated the over-stringency of the axioms."

The conclusion of Lance and Williams (1967) that 'nearest neighbour sorting be regarded as obsolete,' is however equally as extreme as regarding single linkage as the only acceptable hierarchical clustering method. Because of its mathematical advantages and because it can be used on large data sets, it remains a useful technique in some situations. Moreover methods derived from it in an attempt to overcome chaining (see Chapter 7) may be extremely useful. (Other approaches to defining acceptable clustering procedures from a set of required mathematical properties are given in Hartigan, 1967 and Fisher and Van Ness, 1973).

A notable advantage of both single and complete linkage clustering emphasised by Johnson (1967), is their invariance under *monotonic transformations* of the proximity matrix. This means that the methods will give the same results on other proximity matrices, the elements of which are in the same rank order as the original; only the *ordinal* properties of the similarity or dissimilarity measure are of consequence. Consequently the difficulties generally involved in scaling and combining different variables into a proximity measure become of less concern.

A number of empirical investigations of hierarchical clustering techniques have been performed that are helpful in giving indications of which methods are likely to be most useful in practice. Baker (1974) and Hubert (1974) both produce evidence that complete linkage clustering is less sensitive to particular types of observational errors than single linkage. (Defrays, 1977, describes a very efficient algorithm for complete link clustering). Cunningham and Ogilvie (1972) compare seven hierarchical techniques and find that group average clustering performs most satisfactorily overall for the data sets considered. In addition, however, they find that there is a strong interaction in the results between type of input data and the particular clustering method used. Kuiper and Fisher (1975) investigate six hierarchical techniques and find that, for equal numbers of points from multivariate normal distributions, Ward's method classifies almost as well as Fisher's linear discriminant function, *knowing* all the parameters. With unequal sample sizes however, centroid, group average and furthest neighbour are more successful. Blashfield (1976) comes to similar conclusions. Applying four hierarchical clustering methods to data generated from multivariate normal mixtures (see Chapter 6), he finds the following levels of agreement between cluster solutions and actual structure, agreement being quantified by the kappa statistic (see Cohen, 1960).

Method	Median kappa	Interquartile range of kappa
Single linkage	0.06	0.034–0.10
Complete linkage	0.42	0.22–0.58
Group average	0.17	0.06–0.46
Ward's method	0.77	0.42–0.94

Clearly for the data sets considered single linkage performs very poorly and Ward's method very well.

A more comprehensive study by Milligan (1980) showed clearly that no single method could be claimed superior for all types of data. In the presence of outliers, for example, Milligan found that single link, centroid and median clustering were virtually unaffected. Ward's method and group average however performed poorly. In contrast when the data were such that they contained a true cluster structure masked by the addition of 'noise', single linkage, centroid and the median method gave poor results, and Ward's method and group average had the superior performance.

A study by Hands and Everitt (1987) using binary data found that Ward's method performed very well when the data contained approximately equally sized clusters, but poorly when the clusters were of different sizes. In the latter situation centroid clustering appeared to give the most satisfactory results.

4.5 Global fit of a hierarchical clustering solution

Hierarchical clustering techniques impose a hierarchical structure on data and it is usually necessary to consider whether this is merited or whether it introduces unacceptable distortions of the original relationships amongst the individuals, as implied by their observed proximities. The method most commonly used for assessing the match between the dendrogram and the proximity matrix is the *cophenetic correlation coefficient*. This is simply the product moment correlation of the $n(n-1)/2$ entries in the lower half of the observed proximity matrix and the corresponding entries in the so called *cophenetic matrix*, C, the elements of which, c_{ij}, are defined to be the first level in the dendrogram at which individuals i and j occur in the same cluster. Since these values satisfy the ultrametric inequality (see above), the match between dendrogram and data cannot be perfect unless the entries in the proximity matrix are also ultrametric, a situation which seldom occurs in practice.

To illustrate the use of the cophenetic correlation coefficient the data given in Section 4.2.1 may be used. The elements of D and C to be correlated are as follows:

d_{ij}:	2.0	5.0	10.0	9.0	4.0	9.0	8.0	5.0	3.0
c_{ij}:	2.0	5.0	5.0	5.0	4.0	5.0	5.0	4.0	3.0

The cophenetic correlation coefficient takes the value 0.82. Rohlf and Fisher (1968) studied the distribution of this measure under the hypothesis that the individuals are randomly chosen from a single multivariate normal distribution. They found that the average value of the coefficient tends to decrease with n and to be almost independent of the number of variables. They also suggested that values of the cophenetic correlation above 0.8 were usually sufficient to reject the null hypothesis. In a later paper however, Rohlf (1970) warns that 'even a cophenetic correlation near 0.9 does not guarantee that the dendrogram serves as a sufficiently good summary of the phenetic relationships.'

(A number of authors have suggested hierarchical clustering techniques which attempt to minimize a measure of discrepancy between the observed dissimilarities and the fitted ultrametric distances. Examples are the methods described in Hartigan, 1967, Chandon *et al.*, 1980, and De Soete, 1984).

4.6   Partitions from a hierarchy—the number of groups problem

It is often the case, when hierarchical clustering techniques are used in practice, that the investigator is not interested in the complete hierarchy but only in one or two partitions obtained from it. In hierarchical clustering, partitions are achieved by 'cutting' a dendrogram or selecting one of the solutions in the nested sequence of clusterings that comprise the hierarchy. In particular applications it will be of interest to try and determine which of all possible partitions produce the best fit to the data; essentially this means deciding on the appropriate number of clusters for the data. One informal method which is often used for this purpose is to examine the differences between fusion levels in the dendrogram. Large changes are taken to indicate a particular number of clusters. Consider, for example, the dendrogram shown in Figure 4.15. This shows a large difference in the level between two groups and the final stage at which all individuals are in a single group. This would be taken as evidence for considering the two group solution as most relevant. Although this procedure is commonly used and can be helpful, it does carry with it the distinct possibility of influence from *a priori* expectations.

More formal approaches to the number of clusters problem in the context of hierarchical clustering have been suggested by several authors. Duda and Hart (1973), for example, proposed a ratio criterion, $E(2)/E(1)$, where $E(2)$ is the sum of squared errors (see Section 4.2.6) within clusters when the data are partitioned into two clusters, and $E(1)$ gives the squared errors when only one cluster is present. The hypothesis of a single cluster is rejected if the ratio is smaller than a specified critical value.

Calinski and Harabasz (1974) also suggest an index for number of groups based on sum of squares terms, namely

$$\frac{\text{trace}(\mathbf{B})/(g-1)}{\text{trace}(\mathbf{W})/(n-g)} \tag{4.17}$$

where $\mathbf{B}$ and $\mathbf{W}$ are the between and within cluster sum of squares and cross products matrices, and g is the number of groups. The maximum value of the index in the hierarchy is taken to indicate the correct number of groups.

Mojena (1977) suggests a procedure based upon the relative sizes of the different fusion levels in the dendrogram. In detail the proposal is to select the

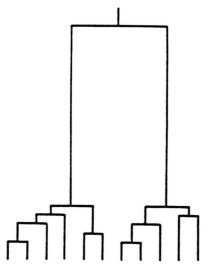

Figure 4.15   Dendrogram indicating two groups.

number of groups corresponding to the first stage in the dendrogram satisfying

$$\alpha_{j+1} > \bar{\alpha} + k s_\alpha \qquad (4.18)$$

where $\alpha_0, \alpha_1, \ldots, \alpha_{n-1}$ are the fusion levels corresponding to stages with $n, n-1$, ..., 1 clusters. The terms $\bar{\alpha}$ and s_α are, respectively, the mean and unbiased standard deviation of the α values and k is a constant. Mojena suggests that values of k in the range 2.75 to 3.50 give the best results.

Milligan and Cooper (1985) report a detailed investigation of indices for number of groups and find that the three mentioned above are amongst the most satisfactory, although they suggest that the value of k for Mojena's stopping rule should be 1.25. Some examples of the use of these indices are given in the next section.

A discussion of whether a data set shows *any* evidence of clustering is left until Chapter 8.

4.7   Applications of agglomerative techniques

In this section several applications of the methods discussed in previous sections are described beginning with a small example taken from Michener (1970).

4.7.1   Classification of forms of the bee *Hoplites producta*

Table 4.1 shows a matrix of distances for eleven forms of the bee *Hoplites producta* based on 23 variables used to describe each form (see Michener, 1970, for details). In an investigation of the structure in such a distance matrix several methods might be applied apart from some form of clustering. Here it might

Table 4.1 Matrix of distance coefficient (based on standardized data) for the forms of the Hoplites production complex (Michener, 1970)

	1	2	3	4	5	6	7	8	9	10	11
1	0										
2	0·940	0									
3	1·229	0·791	0								
4	1·266	0·847	0·303	0							
5	1·507	1·331	1·070	1·026	0						
6	1·609	1·306	0·778	0·573	1·175	0					
7	1·450	1·266	1·475	1·506	1·829	1·876	0				
8	1·239	1·286	1·510	1·540	1·908	1·832	1·655	0			
9	1·493	1·160	0·848	0·792	0·965	0·978	1·847	1·761	0		
10	1·494	1·396	1·497	1·528	1·724	1·687	1·954	1·733	1·721	0	
11	1·348	1·238	1·352	1·385	1·724	1·559	1·844	1·608	1·596	0·645	0

The names of the forms of *Hoplites* are 1, *Hoplites gracilis*; 2, *subgracilis*; 3, *interior*, 4, *bernardina*; 5, *panamintana*; 6, *producta*; 7, *colei*; 8, *elongata*; 9, *uvularis*; 10, *grinelli*; 11, *septentrionalis*.

be useful to produce first an *ordination* of the distances using multidimensional scaling (see Chapter 3). A two-dimensional solution produced by this method is shown in Figure 4.16.

Dendrograms resulting from the application of single linkage, complete linkage and group average clustering are shown in Figures 4.17, 4.18 and 4.19 respectively. 'Cutting' each dendrogram at the stage corresponding to a three group partition, gives the following solutions:

Single linkage: {3, 4, 6, 2, 9, 1, 5, 10, 11}, {7}, {8}

Complete linkage: {3, 4, 6, 5, 9}, {10, 11}, {1, 2, 8, 7}

Group average: {3, 4, 6, 9, 5, 1, 2, 10, 11}, {7}, {8}

Comparing these with the two-dimensional representation of the distances provided by Figure 4.16, suggests perhaps that the solution provided by complete linkage is closest to a 'visual' clustering of the data. It must be remembered however that the fit of the points in Figure 4.16 to the original distance matrix will not be perfect, and that consequently structure inferred from the diagram will only approximate that in the original data. The example serves to illustrate that even on a small data set, different hierarchical clustering procedures may give different solutions.

This example can be used to illustrate Mojena's 'stopping rule'. For complete linkage, $\bar{\alpha} = 0.811$ and $s_\alpha = 0.888$. No stage in the fusion process satisfies the rule if the values of k suggested by Mojena are used. If, however, the revised proposal of Milligan and Cooper, i.e. $k = 1.25$, is employed then the two group solution is accepted. With group average, $\bar{\alpha} = 0.503$ and $s_\alpha = 0.566$. Using again $k = 1.25$, leads to stopping at three groups.

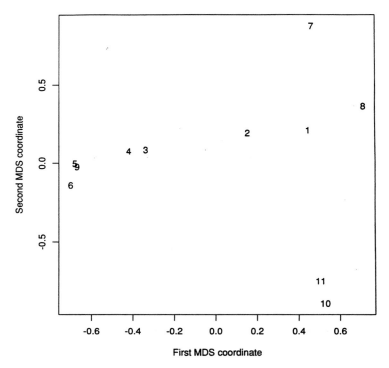

Figure 4.16   Multidimensional scaling solution for distance matrix of species of bee.

4.7.2   Classification of countries on the basis of their athletic prowess

In Chapter 2, Table 2.3 shows the record times for eight running events in 55 countries. These data will be used to classify the countries in terms of their athletic prowess. Here complete linkage is used, first on the Euclidean distances between countries, and then for comparison on the corresponding city block distances. In both cases the data were standardised before calculating distances, since in the raw data some times are given in seconds and some in minutes.

The resulting dendrograms are shown in Figures 4.20 and 4.21. Application of Mojena's stopping rule with $k = 1.25$, leads to selecting the three group solution in both cases. The details of the three group solution are:

Group 1 Countries
    1, 2, 3, 4, 6, 8, 9, 10, 11, 14, 15, 17, 18, 19, 20, 21, 22, 28, 29, 30, 31, 32, 34, 37, 38, 39, 40, 43, 44, 45, 47, 48, 49, 52, 53, 54.

	100m	200m	400m	800m	1500m	5000m	10000m	Marathon
Mean	10.3	20.7	45.9	1.8	3.6	13.4	28.0	131.9

Group 2 Countries
    5, 7, 13, 16, 23, 26, 33, 35, 36, 41, 42, 46, 50, 51.

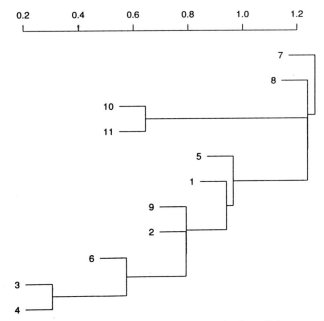

Figure 4.17 Single linkage dendrogram for bee distance matrix.

	100m	200m	400m	800m	1500m	5000m	10000m	Marathon
Mean	10.6	21.4	47.4	1.8	3.8	14.6	30.8	145.9

Group 3 Countries
   12, 55.

	100m	200m	400m	800m	1500m	5000m	10000m	Marathon
Mean	10.8	21.6	47.8	1.9	3.9	14.8	31.3	148.1

Clearly countries 12 (Cook Islands) and 55 (Western Samoa) are considerably different from the remaining 53 countries, which fall into two groups these being the same for each distance measure. The smaller group consists of countries such as Bermuda, Burma, Costa Rica, the Dominican Republic etc., whose athletic ability as reflected by the times for the events considered is less than in those countries in the larger group, for example Argentina, Australia, USA, (former) USSR etc. Here a principal components plot of the data may aid in interpreting results and a plot of the countries in the space of the first two principal components is shown in Figure 4.22(a). In Figure 4.22(b) the three clusters indicated by the cluster analysis are shown. (The first principal component is essentially a measure of total time for all eight events, the second a contrast between sprinting times and long distance times—see Everitt and Dunn, 1991, for details).

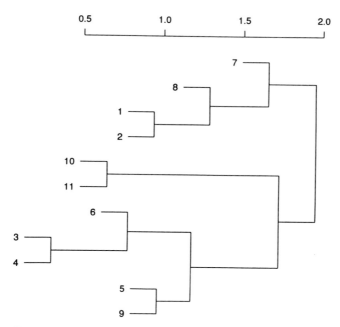

Figure 4.18   Complete linkage dendrogram for bee distance matrix.

4.7.3   The classification of lower back pain

Coste *et al.* (1991) describe an investigation of the clinical and psychological diversity of non-specific low-back pain. One aspect of the study was the use of Ward's clustering technique to produce a classification of 330 subjects complaining of localized low-back pain to hospital rheumatologists. The variables recorded for each subject included information on their medical and surgical history, past and present back symptoms, intensity and duration of pain, type of onset, aggravating and relieving factors and activities. Cluster analysis was applied not to the raw data for each subject but to standardised scores from a principal components analysis—see original paper for details. On the basis of clinical judgement the four class partition produced by Ward's method was deemed most useful. The profiles of these four classes are shown in Table 4.2.

The first cluster, consisting of 46 patients, was characterized by frequent mechanical features (pain increased by movements, impulsion, standing, lifting and relieved by lying; worse pain in the morning). Conversely, non-organic signs (diffuse spinal pain, dysesthesias, increased pain by psychological factors) were very uncommon. The fourth cluster (19 patients) had the opposite characteristics: uncommon mechanical features and physical findings, frequent non-organic signs. The second and the third clusters appeared to be intermediate between the first and fourth. The clusters were further examined in terms of the prevalence of psychiatric disorders, and other cluster analyses performed on subjects with and without psychiatric disorders. Clusters

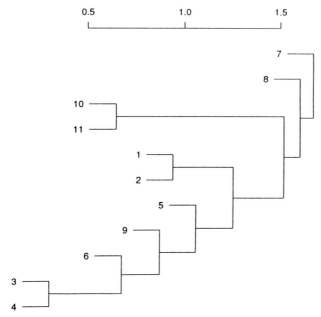

Figure 4.19 Groups average distance matrix for bee distance matrix.

were eventually interpreted through the relationships or interactions between psychological disturbances and the clinical features of low-back pain.

4.7.4 Developing a classification for facial pain

Wastell and Gray (1987) describe the use of clustering for the development of a classification of pain distribution in patients with temporomandibular joint pain dysfunction syndrome (TMJPDS). This refers to a complex symptom group involving facial pain, limitation and deviation of mandibular movements, joint noises and muscle tenderness. Symptoms vary with the stage of the disease; aetiology is equally complex and both physical and psychogenic factors have been implicated. Pain is the most commonly recorded symptom but its facial distribution does not conform to a single pattern.

Wastell and Gray's main aim was to use clustering techniques to develop an objective typology for classifying facial pain in terms of its spatial distribution. Clinically the hope was that the derived classification would be useful in identifying different stages of the disease, which would be of help in defining more directed treatment plans. Data were collected from 127 patients attending the temporomandibular joint clinic of a University hospital complaining of classic TMJPDS. Patients were asked to trace the boundary of their pain-affected area with the tip of their index finger. The examiner recorded this outline on a diagram of the lateral profile of the face (see Figure 4.23) and this was adjusted until patients were satisfied that the outline matched their own pain area.

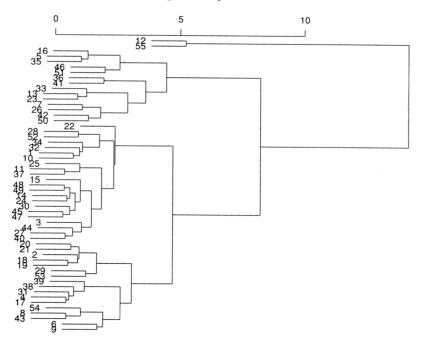

Figure 4.20 Complete linkage dendrogram for athletic records data—Euclidean distance.

The squares of the grid shown in Figure 4.23 falling within the perimeter of a pain area were scored 1; those without, 0. In this way any patient's pain distribution could be described by a string of binary variables. In practice all distributions lay within a central rectangle with horizontal extension J–T and vertical extension 11–28, giving $11 \times 18 = 198$ binary variables for analysis. The similarity matrix calculated from these data used Jacard's coefficient (see Chapter 3), the rationale behind the choice being that in this particular application, it is clearly appropriate to disregard '0–0' matches.

Ward's method of clustering was used and the resulting dendrogram is shown in Figure 4.24. The structure of the dendrogram appears to indicate three major classes, with a possible further subdivision of each of these into two. A composite pain distribution matrix for each class was constructed by simple matrix addition of the pain matrices of its constituent members.

The authors' description of the pain classes is as follows:

Pain Class A: The pain distribution of Class A was concentrated over the temporomandibular joint (see Figure 4.25). The two subclasses were much alike, save for a small vertical difference in their centroids.

Pain Class B: The pain distribution of Class A differed from Class B in involving the vertical portion of the mandible (the ramus). The two subclasses of B were quite different; Class B2 showed a distribution confined to the lower

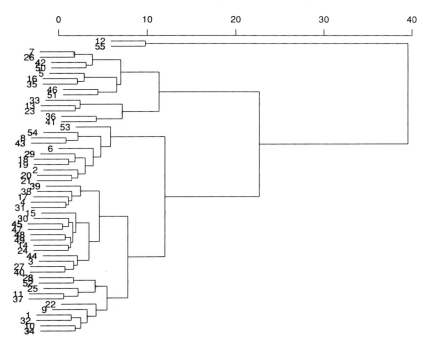

Figure 4.21 Complete linkage dendrogram for athletic records data—city block distance.

part of the ramus and Class B1 a much wider distribution covering all the ramus and the interior part of the temple (see Figure 4.25).

Pain Class C: The pain distribution of Class C differed from the other two classes in involving an anterior projection over the zygomatic arch. The two subclasses were again distinct: Class C1 showed a distribution confined to the temporomandibular joint and the zygomatic arch; Class C2 showed a much wider distribution spreading over the temple as well as covering the ramus (Figure 4.25).

The authors went to much effort to validate their proposed classification using a variety of methods, one of which was to relate the clusters to clinical features. Many such features showed large differences in the derived clusters. The final conclusion was that the groups could be interpreted in terms of a chronological model of the development of TMJPDS.

4.7.5 Trace metals in the brain

A further interesting medical example is provided by Duflou, Maenhaut and De Reuck (1990). Here the concentrations of various trace metals in 46 brain structures taken from individuals who had died from a variety of causes were determined. Both principal components analysis and a variety of hierarchical clustering methods were used to examine the structure of the data. The solutions produced by complete linkage and Ward's method were very sim-

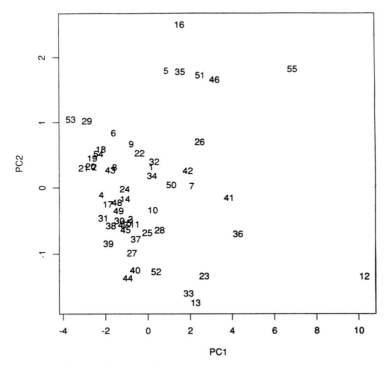

Figure 4.22 (a)   Plot of athletic records data in space of first two principal components. Individual countries shown.

ilar, both consisting of two large clusters with thirty and sixteen structures respectively. The first cluster included all grey matter structures, while the second comprised all white matter. The results indicated a distinct difference in trace element composition between these two groups of brain regions. It was also found that structures involved in the same physiological function or morphologically similar regions often conglomerated in the same subcluster, suggesting a relation between the trace element profile of a brain structure and its function or morphology.

Other applications of hierarchical clustering methods are described in Paykel and Rassaby (1978), Pritchard and Anderson (1971), Murtagh (1985) and Adamson and Bawden (1981).

4.8   Divisive methods

Divisive clustering techniques are essentially of two types, *monothetic* which divide the data on the basis of the possession or otherwise of a single specified attribute, and *polythetic*, where divisions are based on the values taken by all attributes. This class of clustering methods is far less popular than the hierarchical techniques and, consequently, the description given here will be

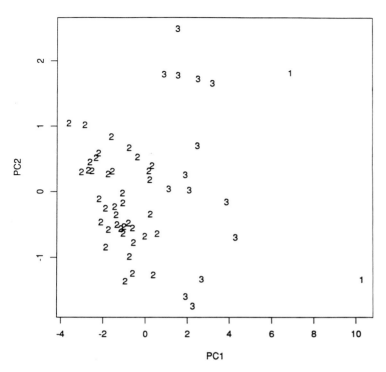

Figure 4.22 (b) Plot of athletic records data in space of first two principal components. Three group solution given by complete linkage identified.

correspondingly brief, concentrating on an example of the polythetic approach and one which involves a monothetic method.

The most feasible of the polythetic divisive methods is that described by MacNaughton-Smith *et al.* (1964). Here a 'splinter' group is accumulated by sequential addition of the individual whose total dissimilarity with the remainder, less its total dissimilarity with the splinter group, is a maximum. When this difference becomes negative the process is repeated on the two sub-groups. The usual measure of dissimilarity used is the average Euclidean distance between each individual and the other individuals in the group. Consider, for example, the following distance matrix, **D**, for seven individuals.

$$
\mathbf{D} = \begin{array}{c} 1 \\ 2 \\ 3 \\ 4 \\ 5 \\ 6 \\ 7 \end{array}
\begin{array}{c}
\begin{array}{ccccccc} 1 & 2 & 3 & 4 & 5 & 6 & 7 \end{array} \\
\left(\begin{array}{ccccccc}
0 & & & & & & \\
10 & 0 & & & & & \\
7 & 7 & 0 & & & & \\
30 & 23 & 21 & 0 & & & \\
29 & 25 & 22 & 7 & 0 & & \\
38 & 34 & 31 & 10 & 11 & 0 & \\
42 & 36 & 36 & 13 & 17 & 9 & 0
\end{array}\right)
\end{array}
$$

Table 4.2   Exploratory classification of lower back pain: distribution of selected clinical features across the four classes

Cluster	1	2	3	4	*All*
No. of observations	46	194	70	19	329*
Mean age (yr)	56	46	51	54	49
Sex (% men)	40	53	33	10	45
Compensation status (%)	17	9	13	5	11
Median duration of LBP (weeks)	53	31	68	36	43
Previous separate attacks (> 0) (%)	100	33	97	58	42
Sudden onset (%)	21	30	23	15	26
In morning, back worse (%)	82	62	70	38	65
Permanent pain in night (%)	13	5	6	10	7
Diffuse spinal pain (%)	8	14	10	21	13
Pain increased by impulsion (%)	54	39	35	35	40
Pain worse on moving back (%)	90	70	75	47	72
Pain worse on standing (%)	65	52	60	42	54
Pain worse with changing climate (%)	44	24	35	12	30
Pain worse by psychologic factors (%)	16	26	20	85	27
Dysesthesias in the back (%)	4	10	6	40	10
Structural accentuated lordosis (%)	21	5	15	10	10
Spondylolisthesis hole (%)	17	3	6	5	7
Limited passive movements (%)	63	50	73	31	55
Catch (%)	26	15	13	5	15
Paravertebral muscular contracture (%)	21	14	15	0	15
Straight leg raising < 75° (%)	40	31	35	0	30
DSM-III diagnosis† (%)	26	44	40	58	41

* Cluster analysis on 329 subjects. One subject was excluded because of unknown number of previous separate attacks of LBP.
† This variable did not participate in any way in the clustering process.

The individual used to initiate the splinter group is the one whose average distance from the other individuals is a maximum. This is found to be individual 1, giving the initial groups

$$(1) \text{ and } (2, 3, 4, 5, 6, 7)$$

Next the average distance of each individual in the main group to the individuals in the splinter group is found, followed by the average distance of each individual in the main group to the other individuals in this group. The difference between these two averages is then found. Here this leads to

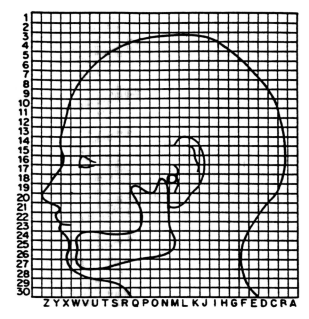

Figure 4.23 TMJPDS-data collection method.

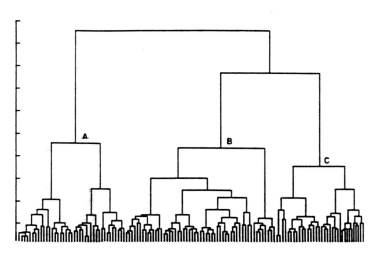

Figure 4.24 Ward's dendrogram on TMJPDS data. (Reproduced with permission from Wastell and Gray, 1987.)

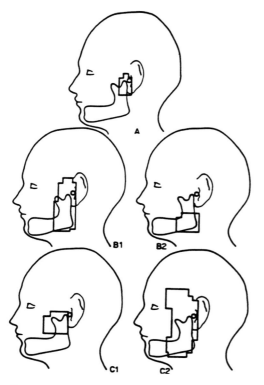

Figure 4.25   Pain distribution of cluster solution on TMJPDS data. (Reproduced with permission from Wastell and Gray, 1987.)

Individual	Average distance to splinter group (1)	Average distance to main group (2)	$(2-1)$
2	10.0	25.0	15.0
3	7.0	23.4	16.4
4	30.0	14.8	−15.2
5	29.0	16.4	−12.6
6	38.0	19.0	−19.0
7	42.0	22.2	−19.8

The maximum difference is 16.4 for individual 3, which is therefore accumulated into the splinter group giving the two groups

$$(1, 3) \text{ and } (2, 4, 5, 6, 7)$$

Repeated, the process gives the following:

Individual	Average distance to splinter group (1)	Average distance to main group (2)	(2 − 1)
2	8.5	29.5	21.0
4	25.5	13.2	−12.3
5	25.5	15.0	−10.5
6	34.5	16.0	−18.5
7	39.0	18.7	−20.3

So now individual 2 joins the splinter group to give groups

$$(1, 3, 2) \text{ and } (4, 5, 6, 7)$$

and the process is repeated to give

Individual	Average distance to splinter group (1)	Average distance to main group (2)	(2 − 1)
4	24.3	10.0	−14.3
5	25.3	11.7	−13.6
6	34.3	10.0	−24.3
7	38.0	13.0	−25.0

As all the differences are now negative, the process would continue (if desired) on each sub-group separately.

This method has the advantage that the computation required is considerably less than for an 'all possible subdivisions' method. As with other divisive techniques, an inefficient early partition cannot be corrected at a later stage, but this is also true of agglomerative methods.

Monothetic techniques are generally used where the data consists of binary variables. A division is then initially into those individuals who possess and those who lack, some one specified attribute. If divisions of this type only are considered then for a data set with m binary variables there are m potential divisions of the initial set, $m - 1$ each of the two sub-sets thus formed and so on. Differences between methods arise because of the different criterion which may be used to choose the particular variable on which to divide. The most common division criteria are based on the chi-square type statistics derived from the fourfold table for each pair of variables, i.e.

$$X_{jk}^2 = \frac{(ad - bc)^2 N}{(a + b)(a + c)(b + d)(c + d)} \tag{4.19}$$

(see Chapter 3).

For example, division might be on that variable, k (say), which makes $\sum_{j \neq k} X_{jk}^2$ a maximum. Each division made can be located at a precise hierarchical level on a division tree according to the max $\sum X^2$ value. Other criteria which have been proposed are

$$\max \sum \sqrt{X_{jk}^2}$$

$$\max \sum |ad - bc| \qquad (4.20)$$

$$\max \sum (ad - bc)^2$$

To illustrate this approach consider the following data set consisting of three binary variables on five individuals.

	Variable		
Individual	1	2	3
1	0	1	1
2	1	1	0
3	1	1	1
4	1	1	0
5	0	0	1

The three chi-square statistics for each pair of variables can be calculated to be $X_{12}^2 = 1.87$, $X_{13}^2 = 2.22$, $X_{23}^2 = 0.83$, giving

$$X_{12}^2 + X_{13}^2 = 4.09$$
$$X_{12}^2 + X_{23}^2 = 2.70$$
$$X_{13}^2 + X_{23}^2 = 3.05$$

Using the $\max \sum X^2$ criterion, then the first division of the data into two sub-sets is into those individuals who possess variable 1 and those who do not, giving the division

$$(2, 3, 4) \text{ and } (1, 5).$$

Such methods have had their widest use in ecological studies; see, for example, Lambert and Williams (1962, 1966) and Pielou (1969). Other applications are described in Ballard and Gothfredson (1963) and Wilkins and MacNaughton-Smith (1964).

4.9   Summary

The concept of the hierarchical representation of a data set was developed primarily in biology. The structures output from a hierarchical clustering method resembles the traditional structure of Linnaean taxonomy with its graded sequence of ranks, and specimens grouped into *species* and these groups themselves classified into *genera* etc. Although any numerical taxonomic exercise with biological data need not necessarily replicate the structure of such a traditional classification, there nevertheless remains a strong tendency amongst biologists for hierarchical classifications. Hierarchical clustering procedures are however now used in many other fields in which hierarchical structures may not be the most appropriate, and the logic of their use in such areas needs careful thought. The danger of *imposing* a hierarchical scheme on data which is essentially non-hierarchical is clear.

Of the two types of hierarchical clustering scheme, the agglomerative methods are far more widely used. Amongst this type, empirical studies point to Ward's method, group average and complete linkage as the most useful in practice, although the results are not clear cut, and a large 'method × type of data structure' interaction seems to exist. In many applications researchers will want to select the partition with the 'best' number of groups for a data set from the overall dendrogram. A number of indices have been suggested for this purpose but it remains a difficult problem.

Excellent and comprehensive reviews of hierarchical clustering methods are given in Gordon (1987) and Gordon (1992).

5

Optimization Methods for Cluster Analysis

5.1 Introduction

In this chapter, a class of clustering techniques is considered which produce a partition of the individuals for a particular number of groups, by either minimizing or maximizing some numerical criterion. Such *optimization* techniques differ from the methods described in the previous chapter in not necessarily forming hierarchical classifications of the data. Differences between the methods in this class arise, both because of the variety of clustering criteria that might be selected, and the various optimization algorithms which might be used. In the initial discussion of these methods it is assumed that the number of groups has been fixed by the investigator. Techniques useful for suggesting the 'correct' number of groups are described in Section 5.5.

5.2 Clustering criteria

The basic idea behind the methods to be described in this chapter, is that associated with each partition of the n individuals into the required number of groups, g, is an index, $f(n, g)$, the value of which is indicative of the 'quality' of this particular clustering. For some indices, high values are sought, for others low values (see later). Associating a number with each partition allows them to be compared. Many such clustering criteria have been suggested, but the most commonly used arise from considering the following three matrices which can be calculated from a partition of the data.

$$\mathbf{T} = \frac{1}{n} \sum_{i=1}^{g} \sum_{j=1}^{n_i} (\mathbf{x}_{ij} - \bar{\mathbf{x}})(\mathbf{x}_{ij} - \bar{\mathbf{x}})'$$

$$\mathbf{W} = \frac{1}{n-g} \sum_{i=1}^{g} \sum_{j=1}^{n_i} (\mathbf{x}_{ij} - \bar{\mathbf{x}}_j)(\mathbf{x}_{ij} - \bar{\mathbf{x}}_i)' \tag{5.1}$$

91

$$\mathbf{B} = \sum_{i=1}^{g} n_i(\bar{\mathbf{x}}_i - \bar{\mathbf{x}})(\bar{\mathbf{x}}_i - \bar{\mathbf{x}})'$$

These $p \times p$ matrices (p is the number of variables) represent respectively, total dispersion, within-group dispersion and between-group dispersion, and satisfy the following equation:

$$\mathbf{T} = \mathbf{W} + \mathbf{B} \tag{5.2}$$

For $p = 1$ this equation represents a relationship between scalars; simply the division of total sum-of-squares for a variable, into the within and between groups sum-of-squares, familiar from a one-way analysis of variance. In this case a natural criterion for grouping would be to choose the partition corresponding to the minimum value of the within group sum-of-squares, or, equivalently, the maximum value of the between group term.

For $p > 1$ the derivation of clustering criteria from equation (5.2) is not so clear cut, and several alternatives have been suggested.

5.2.1   Minimization of trace (W)

An obvious extension of the minimization of the within group sum-of-squares criterion suggested above for the case $p = 1$, when the data are not univariate, is to minimize the sum of the within groups sum-of-squares, over *all* the variables, that is to mimimize trace $(\mathbf{W})$. This can be shown to be equivalent to minimizing the sum of squared Euclidean distances between individuals and their cluster mean, that is

$$E = \sum_{i=1}^{n} d_{i,c(i)}^2 \tag{5.3}$$

where $d_{i,c(i)}$ is the Euclidean distance from individual i, to the mean of the cluster to which it is assigned. (Minimizing trace $(\mathbf{W})$ is, of course, also equivalent to maximizing trace $(\mathbf{B})$).

This criterion was explicitly suggested by Singleton and Kautz (1965), and is implicit in the clustering methods described by Forgey (1965), Jancey (1966), MacQueen (1967) and Ball and Hall (1967). Ward's method, described in the previous chapter, also involves this criterion, but in the framework of hierarchical clustering where only the hierarchical building process is optimised.

5.2.2   Minimization of the determinant of (W)

In multivariate analysis of variance, one of the tests for differences in group mean vectors is based on the ratio of the determinants of the within and total dispersion matrices (see Krzanowski, 1988). Large values of $\det(\mathbf{T})/\det(\mathbf{W})$ indicate that the group mean vectors do differ. Such considerations led Friedman and Rubin (1967) to suggest as a clustering criterion, the maximization of this ratio. Since for all partitions of the n individuals into g groups, $\mathbf{T}$ remains the same, maximization of $\det(\mathbf{T})/\det(\mathbf{W})$ is equivalent to the minimization of $\det(\mathbf{W})$. This criterion has been studied in detail by Marriott (1971, 1982).

5.2.3 Maximization of trace $(\mathbf{BW}^{-1})$

A further criterion suggested by Friedman and Rubin (1967) is the maximization of the trace of the matrix obtained from the product of the between-groups dispersion matrix and the inverse of the within-groups matrix. This function is also used in the context of the multivariate analysis of variance, and is equivalent to what Rao (1952) calls the generalization of Mahalanobis distance to more than two groups. The matrix $\mathbf{BW}^{-1}$ is also encountered in canonical variate analysis.

Both trace$(\mathbf{BW}^{-1})$ and $\det(\mathbf{T})/\det(\mathbf{W})$ may be expressed in terms of the eigenvalues, λ_i, of $\mathbf{BW}^{-1}$, that is

$$\text{trace}(\mathbf{BW}^{-1}) = \sum_{i=1}^{p} \lambda_i \tag{5.4}$$

$$\frac{\det(\mathbf{T})}{\det(\mathbf{W})} = \prod_{i=1}^{p}(1 + \lambda_i) \tag{5.5}$$

Other optimization clustering methods, not based on functions of the sample dispersion matrices, are described in Rubin (1967) and Wallace and Boulton (1968).

All the criteria mentioned thus far are essentially most suitable for data with continuous variables. Spath (1985) describes a number of other criteria suitable for binary and ordinal variables.

5.3 Optimizing the clustering criterion

Once a suitable numerical clustering criterion has been selected, consideration needs to be given to how to choose the g group partition of the data which leads to its optimization. In theory of course the problem is simple; to quote Dr. Idnozo Hcahscror-Tenib, that super galactian hypermetrician who appeared in Thorndike's 1953 presidential address to the Psychometrika Society, 'Is easy. Finite number of combinations. Only, 563 billion billion billion. Try all keep best.'

Unfortunately (and despite the intervening forty years), the problem in practice is not so straightforward. Even with today's computers the numbers involved are so vast, that complete enumeration of *every* possible partition of n individuals into g groups is simply not possible. Some examples taken from Spath (1980) will serve to illustrate the scale of the problem.

$$N(15, 3) = 2,375,101$$
$$N(20, 4) = 45,232,115,901$$
$$N(25, 8) = 690,223,721,118,368,580$$
$$N(100, 5) = 10^{68}$$

where $N(n, g)$ is the number of distinct partitions of n individuals into g non-empty groups. A general expression for $N(n, g)$ is given by Liu (1968):

$$N(n, g) = \frac{1}{g!} \sum_{i=0}^{g} (-1)^{g-i} \binom{g}{i} i^n \tag{5.6}$$

The impracticability of examining every possible partition has led to the development of algorithms designed to search for the optimum value of a clustering criterion by rearranging existing partitions and keeping the new one only if it provides an improvement; these are the so called *hill-climbing* algorithms although in the case of criteria which require minimization they should perhaps be termed *hill descending*. The essential steps in these algorithms are

(a)   Find some initial partition of the individuals into the required number of groups.

(b)   Calculate the change in the clustering criterion produced by moving each individual from its own to another cluster.

(c)   Make the change which leads to the greatest improvement in the value of the clustering criterion.

(d)   Repeat steps (b) and (c) until no move of a single individual causes the clustering criterion to improve.

   (There are, of course, differences in detail between the steps given above, and the exact procedure used in a particular implementation of an optimization cluster method; but (a) to (d) capture the essence of the approach).
   Choosing where to begin the process is achieved in a variety of ways. An initial cluster configuration might be specified on the basis of prior knowledge; it might be the result of some other type of clustering method, for example one of the hierarchical techniques described in the previous chapter. An initial partition might be chosen at random, or g points in the p-dimensional space might be selected in some way to act as initial cluster centres. All these methods have been used at some time or another—for details see MacQueen (1967), Beale (1969a,b), Thorndike (1953), McRae (1971), Friedman and Rubin (1967) and Blashfield (1976). The last author finds, not surprisingly, that the results from an optimization method can, in some cases, be radically affected by the choice of the starting partition. Different initial solutions may lead to a different *local* optimum of the clustering criterion, although with well structured data it is reasonable to expect convergence to the same, hopefully global, optimum from most starting configurations. (Hartigan, 1975, considers empirical and analytic results connecting local and global optima). Marriott (1982) suggests that slow convergence and widely different groupings given by different initial partitions usually indicates that g is wrongly chosen, in particular that there is no evidence of clustering.
   Marriott (1982) also gives a number of results which make it easy to calculate the changes produced in some of the clustering criteria discussed previously when an individual is added to a group. He shows, for example, that the addition of an individual with vector of variable values, $\mathbf{x}$, to group s with

mean $\bar{x}_s$ and size, n_s, changes $\mathbf{W}_s$ to $\mathbf{W}_s^* = \mathbf{W}_s + \mathbf{d}_s\mathbf{d}_s'$ where

$$\mathbf{d}_s = (\mathbf{x} - \bar{\mathbf{x}}_s)\left(\frac{n_s}{n_s + 1}\right)^{\frac{1}{2}} \tag{5.7}$$

Consequently

$$\text{trace}(\mathbf{W}_s^*) = \text{trace}(\mathbf{W}_s) + \mathbf{d}_s'\mathbf{d}_s \tag{5.8}$$

$$\det(\mathbf{W}_s^*) = \det(\mathbf{W}_s)(1 + \mathbf{d}_s'\mathbf{W}_s^{-1}\mathbf{d}_s) \tag{5.9}$$

Before proceeding to discuss the properties and problems of optimization clustering techniques, it will be helpful to consider a small numerical example of the application of the type of algorithm described in this section.

5.3.1 Numerical example

Consider the following data set consisting of the scores for two variables on each of seven individuals.

Individual	Variable 1	Variable 2
1	1.0	1.0
2	1.5	2.0
3	3.0	4.0
4	5.0	7.0
5	3.5	5.0
6	4.5	5.0
7	3.5	4.5

This set of data is to be clustered into 2 groups using the minimization of trace($\mathbf{W}$) method. As a first step in finding a sensible initial partition let the variable values of the two individuals furthest apart (using the Euclidean distance measure), define initial cluster means, giving

	Group 1	Group 2
Individual	1	4
Mean vector	[1.0, 1.0]	[5.0, 7.0]

The remaining individuals are now examined in sequence and allocated to the group to which they are closest, in terms of Euclidean distance, to the cluster mean. The mean vector is recalculated each time a new member is added. This leads to the following series of steps:

	Group 1		Group 2	
	Individual	Mean vector	Individual	Mean vector
Step 1	1	[1.0, 1.0]	4	[5.0, 7.0]
Step 2	1, 2	[1.2, 1.5]	4	[5.0, 7.0]
Step 3	1, 2, 3	[1.8, 2.3]	4	[5.0, 7.0]
Step 4	1, 2, 3	[1.8, 2.3]	4, 5	[4.2, 6.0]
Step 5	1, 2, 3	[1.8, 2.3]	4, 5, 6	[4.3, 5.7]
Step 6	1, 2, 3	[1.8, 2.3]	4, 5, 6, 7	[4.1, 5.4]

This gives the initial classification, the two groups at this stage having the following characteristics:

Group 1 Individuals 1, 2 and 3
    Mean vector $= [1.8, 2.3]$
    trace($\mathbf{W}_1$) $= 6.84$

Group 2 Individuals 4, 5, 6, and 7
    Mean vector $= [4.1, 5.4]$
    trace($\mathbf{W}_2$) $= 5.38$

At this point trace($\mathbf{W}$) $= 6.84 + 5.38 = 12.22$.

Consider now moving individual 3 to the second group, giving trace($\mathbf{W}_1$) $= 0.63$, trace($\mathbf{W}_2$) $= 7.90$ and trace($\mathbf{W}$) $= 8.53$. Since the move causes a decrease in the clustering criterion, the move is made, and the iterative process would now continue from this new partition.

Approaches other than the hill climbing algorithms described above have been suggested to overcome the impossibility of complete enumeration of all possible partitions. Jensen *et al.* (1969), for example, gives details of a dynamic programming algorithm, and Gordon and Henderson (1977) discuss an approach based on the steepest descent technique (see Everitt, 1987). Koontz, Narendra and Fukunaga (1975) use a branch and bound algorithm to search for the global optimum. Such algorithms, whilst generally more efficient than the simple hill climbing approach still suffer from computing time increasing very rapidly with n. A number of improvements to the usual hill climbing algorithms are described by Klein and Dubes (1989) and by Ishmail and Kamel (1989).

An interesting alternative approach to searching for the partition corresponding to min trace($\mathbf{W}$) is provided by Calinski and Harabasz (1974). Starting from the Euclidean distance matrix for pairs of individuals, the minimum spanning tree is found—see Chapter 2. This is then partitioned by removing some of its edges, $(g - 1)$ if a g group solution is sought. The sum-of-squares criterion is then calculated for each of the $\binom{n-1}{g-1}$ possible splits.

5.4 Properties of, and problems with, optimization clustering techniques

Several numerical criteria for clustering have been discussed in the previous sections. Perhaps the most commonly used is minimization of trace($\mathbf{W}$), despite it being well known that it suffers from a number of serious problems. Firstly the method is scale dependent. Different solutions may be obtained from the raw data and from the data standardized in some particular way. That this is so, should be clear from the criterion's equivalent definition in terms of Euclidean distances, and the effect of standardization on the latter (see Chapter 3). Clearly this is of considerable practical importance because of the need for standardization in many applications. A further problem with the use of this criterion is that it may impose a 'spherical' structure on the clusters found even when the 'natural' clusters in the data are of other shapes. As an

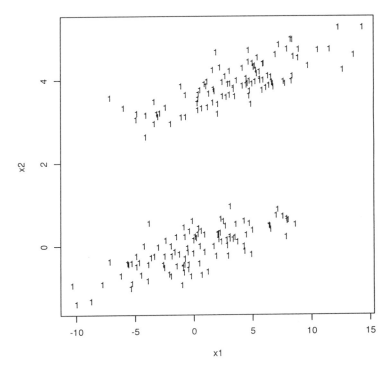

Figure 5.1   Two well separated elliptical clusters

example consider the data shown in Figure 5.1, containing two well separated 'elliptical' clusters. (These data were generated as detailed in Chapter 4, page 69). The two group solution given by the minimization of trace(**W**) applied to these data is shown in Figure 5.2. Clearly the 'wrong' structure has been forced upon the data.

The scale dependency of the trace(**W**) method was the motivation behind Friedman and Rubin's (1967) search for alternative criteria *not* affected by scaling. Of their suggestions discussed earlier in this chapter, that based on minimizing det(**W**), has been most widely used. Scott and Symons (1971) show how the criterion arises from likelihood ratio considerations and Binder (1978), using a Bayesian approach to clustering, shows that it may be further justified as maximizing certain approximate posterior probabilities. Spath (1980) gives a nice illustration of its lack of scale dependency in a comparison with the trace(**W**) criterion—see Figure 5.3.

Apart from its advantages with regard to standardization, the det(**W**) criterion has a further point in its favour, namely that it does not restrict clusters to being spherical. This is clearly seen in its two group solution for the data shown in Figure 5.1—see Figure 5.4. Unfortunately the det(**W**) criterion does assume that all the clusters in the data have the same shape, and again this can cause problems when the data do not conform to the requirement. As

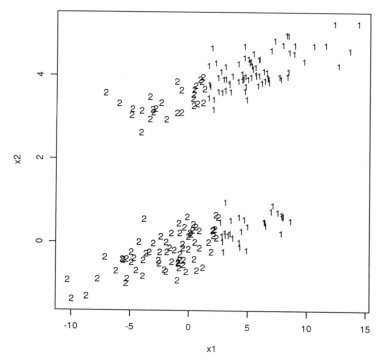

Figure 5.2 Two group solution given by minimization of trace(**W**) applied to the data shown in Figure 5.1.

an example consider the data shown in Figure 5.5. The two group solution given by minimizing det(**W**) is shown in Figure 5.6. Some observations clearly belonging in the elliptical group are misplaced because of the desire to find clusters of similar shape.

In an attempt to overcome the 'similar shape' problem Scott and Symons (1971) suggest a clustering method based on the minimization of

$$\prod_{i=1}^{g} \det(\mathbf{W}_i)^{n_i}.$$

(The restriction that each cluster contains at least $p+1$ individuals is necessary to avoid singular within group dispersion matrices, the determinants of which would be zero). An alternative possibility described by Maronna and Jacovkis (1974) is the minimization of

$$\sum_{i=1}^{g} (n_i - 1)\det(\mathbf{W}_i)^{\frac{1}{p}}.$$

As an illustration of how such criteria may perform more successfully than

2 CLUSTERS

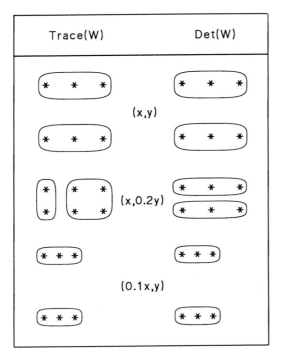

Figure 5.3 Illustration of scale dependency of minimization of trace (**W**) compared to minimization of det(**W**). (Reproduced with permission from Spath, 1980.)

the simpler det(**W**) criterion, the method suggested by Scott and Symons was applied to the data shown in Figure 5.5, and resulted in the correct groups being recovered.

The tendency for criteria such as trace(**W**) and det(**W**) to give equal sized groups has been commented on by a number of authors—for example, Scott and Symons (1971). (An illustration of the problem is given in Figure 5.7). In an attempt to overcome this problem, Symons (1981) suggested two further clustering criteria

$$n \ln \det(\mathbf{W}) - 2 \sum n_i \ln n_i \qquad (5.10)$$

$$\sum (n_i \ln \det \mathbf{W}_i - 2n_i \ln n_i) \qquad (5.11)$$

These and the earlier suggestions of Scott and Symons (1971) and Maronna and Jacovkis (1974) are all considered by Marriott (1982) in terms of the change produced by adding an individual to a cluster. He concludes that the criteria of Symons may have several desirable properties and that these and the other suggestions that have been made are worthy of further study.

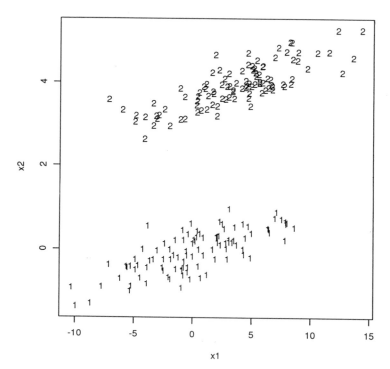

Figure 5.4 Two group solution give by minimization of det(**W**) applied to data shown in Figure 5.1.

5.5   Selecting the number of groups

In most applications of optimization methods of cluster analysis, the investigator will have to 'estimate' the number of clusters in the data. A variety of methods have been suggested which may be helpful in particular situations. Most are relatively informal and involve, essentially, plotting the value of the clustering criterion against the number of groups. Large changes of level in the plot are usually taken as suggestive of a particular number of groups. Like similar procedures for judging dendrograms (see Chapter 4) this approach may be very subjective.

A number of more formal techniques have been described which try to overcome the problem of subjectivity. Beale (1969), for example, gives an 'F test', which may be used to test whether a sub-division into g_2 clusters is significantly better than a subdivision into some smaller number of clusters, g_1. The test statistic is defined as follows:

$$F(g_1, g_2) = \frac{R_{g_1} - R_{g_2}}{R_{g_2}} \bigg/ \left[\left\{ \frac{n - g_1}{n - g_2} \right\} \left(\frac{g_2}{g_1} \right)^{2/p} - 1 \right] \qquad (5.12)$$

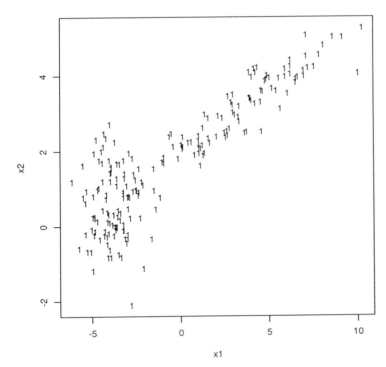

Figure 5.5 Two well separated clusters of different shapes

where $R_g = (n - g)S_g^2$ and S_g^2 is the mean square deviation from cluster centres in the sample. The statistic is compared with F, with $p(g_2 - g_1)$ and $p(n - g_2)$ d.f. A significant result indicates that a subdivision into g_2 clusters is an improvement over a subdivision into the smaller number, g_1. Experience with this procedure suggests that it will only be useful when the clusters are fairly well separated and approximately spherical in shape.

A method suggested by Calinski and Harabasz (1974) is to take the value of g which corresponds to the maximum value of C, where C is given by

$$C = \frac{\text{trace}(\mathbf{B})}{g - 1} \Big/ \frac{\text{trace}(\mathbf{W})}{n - g} \qquad (5.13)$$

This criterion performs reasonably well in the study of indicators of number of groups reported in Milligan and Cooper (1985).

Marriott (1971) suggests as a possible procedure for assessing the number of groups, taking the value of g for which $g^2 \det(\mathbf{W})$ is a minimum. For unimodal distributions, Marriott shows that this is likely to lead to accepting that $g = 1$, and for strongly grouped data it will lead to the appropriate value of g. The sampling properties of the associated statistic, $g^2 \det(\mathbf{W})/\det(\mathbf{T})$, under the assumption that the population has a uniform distribution are investigated by

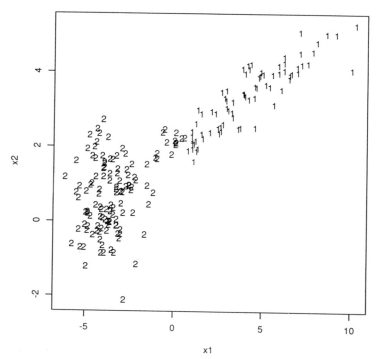

Figure 5.6 Two group solution given by minimization of det(**W**) applied to data shown in Figure 5.5.

Monte Carlo methods, and the use of these results may be helpful in reaching a decision over the appropriate number of groups for the data.

Some examples of the use of these methods will be given in the next section.

5.6   Applications of optimization methods

In this section several applications of optimization type clustering methods are discussed, beginning with a data set taken from Davenport and Studdert-Kennedy (1972).

5.6.1   Aesthetic judgements on painters

A seventeenth-century critic, Roger de Piles, expressed in quantitative terms a series of aesthetic judgements on 56 painters, using four standard but complex conceptual judgements. De Piles set out to divide 'the chief parts of the art into four columns to wit *Composition*, *Design*, *Colouring* and *Expression*', and on each dimension scored his 56 painters on a scale between 0 and 20, with the latter score reserved for the 'sovereign perfection, which no man has fully arrived at.' The data are given in Table 5.1. (Two missing values present in

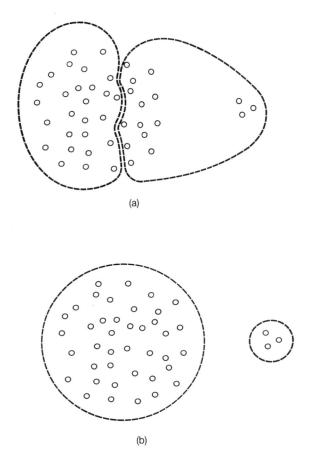

(a)

(b)

Figure 5.7 Illustration of 'equal sized groups' problem of minimization of trace(**W**) clustering.

Davenport and Studdert-Kennedy's listing of these data have been replaced by estimated values from the application of the EM algorithm described in Little and Rubin, 1987).

In an attempt to organize the data in a manner which might shed further light on the nature of de Piles's and, by extension, early eighteenth century artistic judgement, the painters were clustered using the minimization of det(**W**) method. Solutions from two to four groups were found, with, in each case, four random starting configurations being considered. The composition of the groups and the group mean vectors for each solution are given in Table 5.2. Values of $g^2 \det(\mathbf{W})/\det(\mathbf{T})$ were calculated to assess whether a particular number of groups was clearly indicated for these data. The results obtained were

Table 5.1 Artist data (Reproduced with permission from Davenport and Studdert-Kennedy, 1972)

Painter	Composition	Drawing	Colouring	Expression	School
1. Albani	14	14	10	6	e
2. Durer	8	10	10	8	f
3. Del Sarto	12	16	9	8	a
4. Barocci	14	15	6	10	c
5. Bassano	6	8	17	0	d
6. Del Piombo	8	13	16	7	a
7. Bellini	4	6	14	0	d
8. Bourdon	10	8	8	4	h
9. Le Brun	16	16	8	16	h
10. Veronese	15	10	16	3	d
11. The Carracci	15	17	13	13	e
12. Corregio	13	13	15	12	e
13. Volterra	12	15	5	8	b
14. Dipenbeck	11	10	14	6	g
15. Domenichino	15	17	9	17	e
16. Giogione	8	9	18	4	d
17. Guercino	18	10	10	4	e
18. Guido Reni	13·53*	13	9	12	e
19. Holbein	9	10	16	13	f
20. Da Udine	10	8	16	3	a
21. J. Jordaens	10	8	16	6	g
22. L. Jordaens	13	12	9	6	c
23. Josepin	10	10	6	2	c
24. Romano	15	16	4	14	a
25. Lanfranco	14	13	10	5	e
26. Da Vinci	15	16	4	14	a
27. Van Leyden	8	6	6	4	f
28. Michelangelo	8	17	4	8	a
29. Caravaggio	6	6	16	0	e
30. Murillo	6	8	15	4	d
31. Venius	13	14	10	10	g
32. Vecchio	5	6	16	0	d
33. Giovane	12	9	14	6	d
34. Parmigiano	10	15	6	6	b
35. Penni	0	15	8	0	a
36. Perino del Vaga	15	16	7	6	a
37. Cortona	16	14	12	6	c
38. Perugino	4	12	10	4	a
39. Polidore da Cara.	10	17	7·58*	15	a
40. Pordenone	8	14	17	5	d
41. Pourbus	4	15	6	6	f
42. Poussin	15	17	6	15	h
43. Primaticcio	15	14	7	10	b
44. Raphael	17	18	12	18	a
45. Rembrandt	15	6	17	12	g
46. Rubens	18	13	17	17	g
47. Salviata	13	15	8	8	b
48. Le Sueur	15	15	4	15	h
49. Teniers	15	12	13	6	g
50. Testa	11	15	0	6	c
51. Tintoretto	15	14	16	4	d
52. Titian	12	15	18	6	d
53. Van Dyck	15	10	17	13	g
54. Vanius	15	15	12	13	c
55. T. Zuccaro	13	14	10	9	b
56. F. Zuccaro	10	13	8	8	b

* This variable value is missing in the original data. These values were estimated using the EM algorithm described in Little and Rubin (1987).

a = Renaissance, b = Mannerist, c = Seicento, d = Venetian, e = Lombard, f = Sixteenth century, g = Seventeenth century, h = French.

No. of groups(g)	$g^2 \det(\mathbf{W})/\det(\mathbf{T})$
1	1.000
2	0.785
3	0.643
4	0.611

For these data this index for number of groups does not appear to be particularly helpful, although the decreasing values of the index do seem to suggest that the data have *some* structure.

Figure 5.8 displays each of the solutions in the space of the first two principal components of the data.

It is difficult to speculate on these results without being an informed art-historian. One thing which is apparent, however, in the solutions shown in Table 5.2 and in solutions for larger numbers of clusters not given here, is that correspondence between clusters and the school of an artist is relatively small. A further small point is that several of those masters whom de Piles and his contemporaries would have considered as embodying and continuing the 'grand tradition,' are always grouped together—see, for example, le Brun (9), da Vinci (26), Poussin (42) and le Sueur (48).

For these data the solutions given by minimizing trace($\mathbf{W}$) are very similar to those detailed in Table 5.2. See Davenport and Studdert-Kennedy (1972) for details.

5.6.2 Classification of psychiatric patients

Diseases of the mind are more elusive than diseases of the body, and the classification of psychiatric illness has always been a difficult and controversial subject. Clustering techniques have often been used in attempts to refine or even redefine current psychiatric diagnostic systems. One of the earliest of such studies was that of Zubin (1938). Others are those of Lorr *et al.* (1963) and Everitt, Gourlay and Kendell (1971). The latter authors used a k-means algorithm, essentially seeking to minimize trace($\mathbf{W}$), in the clustering of two different sets of psychiatric patients, one from the USA and one from the UK. Each set consisted of 250 patients measured on 45 mental state items, and 25 history items. These measurements were reduced to 10 principal component scores for the cluster analysis. Clusters were found which corresponded largely to standard diagnostic categories such as depression, schizophrenia, and mania, although in each case a large 'ragbag' group was found consisting of patients with very different diagnostic labels.

5.6.3 Classification of 'non-specific' low back pain

The ambiguity of presently available diagnostics for low back pain is, according to Heinrich *et al.* (1985), 'detrimental to the patient's morale and impedes research for optimal treatment and prevention.' Consequently these authors applied several methods of multivariate analysis to a set of 132 signs and symptoms collected on 301 patients suffering from non-specific low back pain, in the search for a useful classification. Amongst these techniques were minimization of trace($\mathbf{W}$) and minimization of det($\mathbf{W}$) clustering. Although

Table 5.2   Clustering of painters: results from minimization of det(**W**)

(1) Two groups

Group 1: n = 35

1, 3, 4, 9, 10, 11, 12, 13, 15, 17, 18, 22, 23, 24, 25, 26, 28, 31, 34, 36, 37, 39, 42, 43, 44, 46, 47, 48, 49, 50, 51, 52, 54, 55, 56

Group 2: n = 21

2, 5, 6, 7, 8, 14, 16, 19, 20, 21, 27, 29, 30, 32, 33, 35, 38, 40, 41, 45, 53

	Composition	Mean Drawing	Colouring	Expression
Group 1	13·7	14·4	9·2	9·6
Group 2	7·9	9·4	13·7	5·0

(2) Three groups

Group 1: n = 13

5, 6, 7, 16, 20, 29, 30, 32, 35, 38, 40, 41, 52

Group 2: n = 27

1, 2, 3, 4, 8, 10, 13, 14, 17, 21, 22, 23, 25, 27, 28, 31, 33, 34, 36, 37, 43, 47, 49, 50, 51, 56

Group 3: n = 16

9, 11, 12, 15, 18, 19, 24, 26, 39, 42, 44, 45, 46, 48, 53, 54

	Composition	Mean Drawing	Colouring	Expression
Group 1	6·2	10·4	14·4	3·0
Group 2	12·4	12·5	9·2	6·4
Group 3	14·5	14·3	10·7	14·3

(3) Four groups

Group 1: n = 16   2, 5, 7, 8, 14, 16, 19, 20, 21, 27, 29, 30, 32, 33, 45, 53
Group 2: n = 15   9, 11, 12, 15, 18, 24, 26, 31, 39, 42, 44, 46, 48, 54, 56
Group 3: n = 18   1, 3, 4, 10, 13, 17, 22, 23, 25, 34, 36, 37, 43, 47, 49, 50, 51, 55
Group 4: n = 7   6, 28, 35, 38, 40, 41, 52

	Composition	Mean Drawing	Colouring	Expression
Group 1	8·9	8·0	14·4	5·2
Group 2	14·4	15·3	9·3	13·9
Group 3	13·6	13·5	8·9	6·3
Group 4	6·3	14·4	11·3	5·1

the results from the different methods were not entirely consistent, five strands of stable group description could be identified.

(1)   A group with patients showing high scores on the general pain indices.
(2)   A group with patients characterised by high scores on the bilateral pain indices.
(3)   A group with patients showing most frequently their pain switching sides.

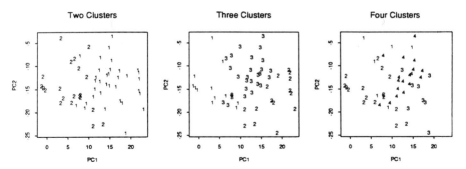

Figure 5.8 Cluster solutions given by minimization of det(**W**) on data in Table 5.1, displayed in the space of the first two principal components.

(4) A group of patients labelled by the absence of signs and symptoms.
(5) A group of patients predominantly showing the presence of anterior disc changes, the absence of reflexes, the presence of sciatica and ipselateral pain in correction with an acute condition.

5.7 Summary

Several optimization clustering techniques have been suggested but two remain most popular—minimization of trace(**W**) and minimization of det(**W**). The latter has considerable advantages over the former, particularly in being invariant to scale changes in the observed variables; additionally it does not make such restrictive assumptions about the shape of clusters. It appears that optimization clustering methods have not found as wide a degree of acceptance as the hierarchical procedures described in the previous chapter. Certainly the number of applications found in the literature is far fewer.

6

Mixture Models for Cluster Analysis

6.1 Introduction

The many methods of cluster analysis described in previous chapters have one feature in common—none is based on the type of model which is generally an essential feature of methods of analysis in other areas of statistics. Consequently the question of inferences from sample to population has not arisen except in an ad hoc fashion. Indeed in most cases the data sets analysed have not been considered as samples from some population of interest at all. For the many users who consider clustering techniques as useful primarily for the exploration of their data, the lack of a statistically acceptable inferential framework is of little consequence. Nevertheless a number of statisticians have expressed concern about this particular aspect of the clustering problem (see, for example, Aitkin, Anderson and Hinde, 1981), and in response have suggested a number of methods based on respectable (if in some cases unrealistic) statistical models. The most common of these approaches is that which uses *mixture distributions*.

6.2 Finite mixture distributions

Suppose a sample of individuals from some population of interest have their height recorded. Such a sample will contain males and females whose average height is known to differ. If the sex of each individual is recorded it is clearly a trivial problem to estimate the mean (and, if required, variance), of men and women. If, however, the individuals are not labelled in this respect the estimation problem is no longer simple. Now the density function of height will have the following form

$$h(\text{height}) = p(\text{female})h_1(\text{height}; \text{female}) + p(\text{male})h_2(\text{height}; \text{male}) \qquad (6.1)$$

109

where p(female) and p(male) are, respectively, the probabilities that a member of the population is a woman or a man, and h_1 and h_2 are the height density functions for women and men. Thus the density function of height has been expressed as a superposition of two conditional density functions; it is known as a *finite mixture density*.

When data are not available for each conditional distribution separately, but only for the overall mixture distribution, the estimation of p(female) and the parameters of the conditional height densities can be difficult; nevertheless it was first considered at the end of the nineteenth century by Karl Pearson, who assumed that the component densities h_1 and h_2 were normal. It is normal mixtures which have become most popular as the basis of a more formal approach to clustering.

6.3   Normal mixtures

Pearson (1894) considered the estimation of the five parameters in the following mixture density

$$f(x) = pf_1(x) + (1-p)f_2(x) \tag{6.2}$$

with

$$f_1(x) = \frac{1}{\sqrt{2\pi\sigma_1^2}} \exp -\frac{(x-\mu_1)^2}{2\sigma_1^2} \tag{6.3}$$

and

$$f_2(x) = \frac{1}{\sqrt{2\pi\sigma_2^2}} \exp -\frac{(x-\mu_2)^2}{2\sigma_2^2} \tag{6.4}$$

Pearson derived estimators by equating sample moments to corresponding population or theoretical moments. The evaluation of his estimators involved the solution of a ninth degree polynomial equation. Nowadays parameter estimates may be found routinely using maximum likelihood methods (see later).

To be a useful model for cluster analysis, the density in (6.2) has to be extended to accommodate more than two components (each component of the mixture being associated with a cluster of observations in the data), and more than a single variable. Such an extension is relatively straightforward,

$$f(\mathbf{x}) = \sum_{i=1}^{g} p_i \alpha(\mathbf{x}, \boldsymbol{\mu}_i, \boldsymbol{\Sigma}_i) \tag{6.5}$$

where

$$0 \leqslant p_i \leqslant 1; \sum_{i=1}^{g} p_i = 1$$

and

$$\alpha(\mathbf{x}, \boldsymbol{\mu}_i, \boldsymbol{\Sigma}_i) = \frac{1}{(2\pi)^{\frac{g}{2}}} |\boldsymbol{\Sigma}_i|^{-\frac{1}{2}} \exp -\frac{1}{2}(\mathbf{x}-\boldsymbol{\mu}_i)'\boldsymbol{\Sigma}_i^{-1}(\mathbf{x}-\boldsymbol{\mu}_i) \tag{6.6}$$

This would be a suitable model when the data are considered to contain g clusters and within each cluster the variables have a *multivariate normal density*, with a particular mean vector and covariance matrix.

Clusters would be formed on the basis of the maximum values of the estimated posterior probabilities

$$\hat{p}(s/\mathbf{x}) = \frac{\hat{p}_s\alpha(\mathbf{x}, \hat{\boldsymbol{\mu}}_s, \hat{\boldsymbol{\Sigma}}_s)}{\sum_{i=1}^{g} \hat{p}_i\alpha(\mathbf{x}, \hat{\boldsymbol{\mu}}_i, \hat{\boldsymbol{\Sigma}}_i)} \tag{6.7}$$

where $\hat{p}(s/\mathbf{x})$ is the estimated probability that an individual with vector of observations, $\mathbf{x}$, belongs to group s. Before these estimated probabilities can be obtained however, consideration has to be given to how the parameters of the density in (6.5) are estimated.

6.3.1   Maximum likelihood estimation

Mixtures of multivariate normal densities have been considered by Cooper (1964), Ihm (1965), Wolfe (1970), Day (1969), Everitt and Hand (1981), Titterington, Smith and Makov (1985) and McLachlan and Basford (1988). The most commom method of estimating parameters has been via maximum likelihood. The log-likelihood function for a sample of observations, $\mathbf{x}_1, \mathbf{x}_2, \ldots, \mathbf{x}_n$ from the mixture density (6.5) is given by

$$L = \sum_{i=1}^{n} \ln \left[\sum_{j=1}^{g} p_j\alpha(\mathbf{x}_i, \boldsymbol{\mu}_j, \boldsymbol{\Sigma}_j) \right] \tag{6.8}$$

The maximum likelihood equations are obtained by setting the first partial derivatives of (6.8) with respect to each parameter to zero. Everitt and Hand (1981) show that this eventually leads to the following series of equations

$$\hat{p}_i = \frac{1}{n} \sum_{j=1}^{n} \hat{p}(i/\mathbf{x}_j), \quad i = 1, 2, \ldots, g - 1 \tag{6.9}$$

$$\hat{\boldsymbol{\mu}}_i = \frac{1}{n\hat{p}_i} \sum_{j=1}^{n} \hat{p}(i/\mathbf{x}_j)\mathbf{x}_j, \quad i = 1, 2, \ldots, g \tag{6.10}$$

$$\hat{\boldsymbol{\Sigma}}_i = \frac{1}{n\hat{p}_i} \sum_{j=1}^{n} \hat{p}(i/\mathbf{x}_j)(\mathbf{x}_j - \hat{\boldsymbol{\mu}}_i)(\mathbf{x}_j - \hat{\boldsymbol{\mu}}_i)', \quad i = 1, 2, \ldots, g \tag{6.11}$$

Written in this form the maximum likelihood equations are seen to be analogous to those for estimating the parameters of a single normal distribution, except that here each sample point is weighted by the posterior probability (6.7). Equations (6.9) to (6.11) do not, of course, give the parameter estimates explicitly; instead they must be solved using some form of iterative procedure. The one usually employed is that first suggested by Hasselblad (1966) and Wolfe (1969). Initial estimates of p_i, $\boldsymbol{\mu}_i$ and $\boldsymbol{\Sigma}_i$ are obtained by one of a variety of methods (see McLachlan, 1992, for suggestions), and these are used to calculate initial values for the posterior probabilities in (6.7). Alternatively,

Table 6.1   Results from fitting two component normal mixture to data in Figure 6.1(a)

	Proportion	Means	SDs	Correlation
Cluster 1	0·50	[1·14, 0·64]	[0·95, 1·10]	0·16
Cluster 2	0·50	[3·94, 4·98]	[2·32, 0·45]	−0·22

initial values of the posterior probabilities may be allocated directly. (Other clustering methods are a possible source of these initial values). Equations (6.9) to (6.11) are then used to find revised parameter values, and the process is continued until some convergence criterion is satisfied. A detailed account of this procedure and its properties is given in Dempster, Laird and Rubin (1977) where it is termed the E (expectation) M (maximization) algorithm.

Problems may arise with maximum likelihood estimation if no constraints are placed on the covariance matrices for each component density. These problems are discussed in detail in Everitt and Hand (1981). A consequence however is that in many applications, these covariance matrices are constrained to be equal.

6.3.2   Numerical example

The bivariate data displayed in Figure 6.1(a) were generated by sampling 50 observations from each of two bivariate normal densities with the following characteristics;

Density one:
$$\mu_x = 1.0, \mu_y = 1.0$$
$$\sigma_x = 1.0, \sigma_y = 1.0$$
$$\rho = 0.0$$

Density two:
$$\mu_x = 4.0, \mu_y = 5.0$$
$$\sigma_x = 2.0, \sigma_y = 0.5$$
$$\rho = 0.0$$

Using the 'true' parameter values as initial estimates, the maximum likelihood procedure outlined in the previous section gave the results shown in Table 6.1. A contour plot of the estimated density function is shown in Figure 6.1(b) and a perspective view of the density in Figure 6.2.

6.3.3   Properties and problems of the normal mixture approach to clustering

A number of authors have investigated the properties of the maximum likelihood estimators for normal mixture distributions, see for example Hosmer (1973), Day (1969), Tan and Chang (1972) and Dick and Bowden (1973). They find that the method will not give accurate estimates unless the component densities are well separated or the sample sizes very large. Convergence to a *local* rather than the *global* maximum of the likelihood function can also be a problem, and in most situations several different sets of initial values need to be used to begin the EM algorithm. In some cases the EM algorithm may be very slow to converge.

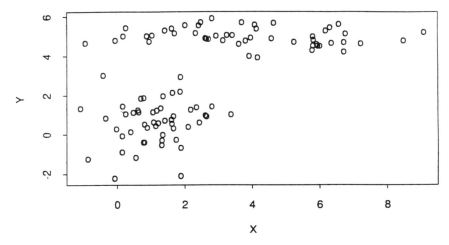

Figure 6.1(a)   Bivariate data containing two clusters.

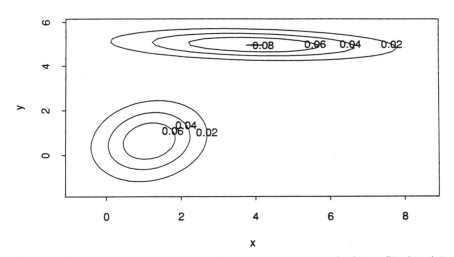

Figure 6.1(b)   Contour plot of estimated two component normal mixture fitted to data in Figure 6.1(a).

An obvious problem with this approach to clustering is the question of whether the model is realistic for the data under investigation. It should be clear however, that this is no more of a problem for this method, where the model used is made explicit, than for those methods where an implicit model at least is involved. Minimizing trace($\mathbf{W}$) for example, essentially involves the assumption of a 'spherical' cluster model.

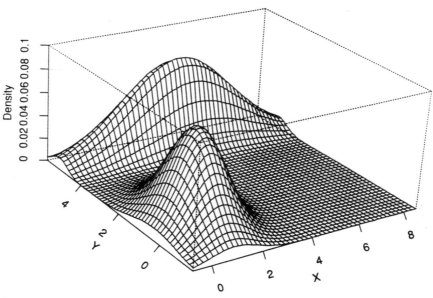

Figure 6.2 Perspective view of estimated two component normal mixture fitted to the data in Figure 6.1(a).

6.4   Estimating g, the number of components in a mixture

According to McLachlan and Basford (1988), 'testing for the number of components, g, in a mixture is an important but very difficult problem which has not been completely resolved.' The obvious way of approaching the problem, using the likelihood ratio test statistic λ to test for the smallest value of g compatible with the data, is, unfortunately, not quite so straightforward with mixture models. Regularity conditions do not hold for $-2\ln\lambda$ to have its usual asymptotic null distribution of chi-squared with degrees of freedom equal to the difference in the number of parameters in the two hypotheses—see McLachlan and Basford (1988) for details.

A number of authors have considered modifications to the likelihood ratio statistic to make it more suitable for testing for the number of components in a mixture. An early proposal, for example, was that of Wolfe (1971), who recommended, on the basis of a small scale simulation study, that the null distribution of $-2\ln\lambda$ for assessing the null hypothesis that the data arise from a mixture of g_1 populations, versus the alternative that they arise from $g_2(g_1 < g_2)$, be approximated as

$$-2c \ln \lambda \sim \chi_d^2 \tag{6.12}$$

where the degrees of freedom, d, are taken as twice the difference in the number of parameters in the two hypotheses, not including the mixing proportions.

Wolfe suggested the following value for c,

$$c = (n - 1 - p - \frac{1}{2}g_2)/n \tag{6.13}$$

On the basis of some simulations for normal populations, Hernadez-Avilia (1979) concluded that Wolfe's proposal was reasonable. More extensive simulations by Everitt (1981) suggest that the ratio of number of observations to number of variables needs to be at least five if Wolfe's approximation is to be applicable for the determination of P values. Everitt also showed that the power of the test is low in cases where the Mahalanobis distance is less than two. Further investigation of the problem is described in Anderson (1985).

As a consequence of these difficulties, McLachlan and Basford (1988) recommend that the outcome of Wolfe's modified likelihood ratio test should not be interpreted too rigidly, but rather used as a guide to the possible number of underlying groups. Examples of the use of the test are given in the next section.

6.5   Applications of normal mixtures

Normal mixture densities have been applied to a wide variety of problems in a number of disciplines. Most common has been their application in the univariate case, and two examples are given in the following section.

6.5.1   Mixtures of normal densities for a single variable

A recent example of a data set modelled by a univariate normal mixture is that given by Roeder (1990). The data set (given here in Table 6.2), concerns the velocities of 82 galaxies from 6 well separated conic sections of space and is intended to shed light on whether or not the observable universe contains superclusters of galaxies surrounded by large voids. The evidence for the existence of the superclusters would be in the multimodality of the distribution of velocities.

A two component normal density with each component assumed to have the same variance was fitted to these data with the results shown in Figure 6.3. There appears to be some evidence of bimodality in the data, although more detailed investigation would be needed before making any definite claims.

As a further example of the fitting of mixture densities to univariate data, blood pressure recordings collected by Clark *et al.* (1968) will be considered. (Histograms for both systolic and diastolic blood pressure are shown in Figure 6.4). Some clinicians, for example, Pickering (1961) claim that blood pressure represents a measurement of a continuous variable from a single population; others such as Platt (1959, 1963) suggest that there are two or three separate sub-populations with different mean levels. The basis of Platt's view is a hypothesis of a dominant gene causing the tendency for arterial blood pressure to rise faster with age for those possessing the gene than it does for persons lacking it.

The results found by Clark *et al.* from fitting two component normal mixtures to various sub-sets of their data are shown in Table 6.3. Attempts were

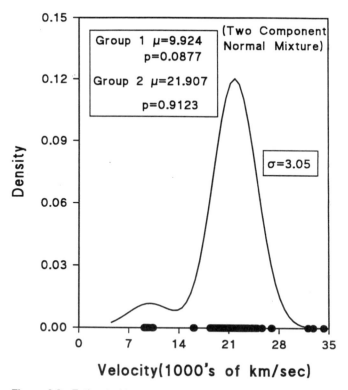

Figure 6.3 Estimated two component mixture fitted to data on velocity of galaxies given in Table 6.2.

made to assess the fit of the estimated mixture densities using chi-square procedures. The final conclusion was that there was little evidence for the sub-group hypothesis in the case of systolic blood pressure, but some rather weak evidence of the existence of two groups in the diastolic data.

6.5.2   Multivariate normal mixtures

The first example of fitting multivariate normal mixture densities involves the Fisher–Anderson iris data (see Chapter 2). The EM algorithm was used to find parameter estimates for two, three and four component solutions. Covariance matrices were not constrained to be equal. The results of Wolfe's modified likelihood ratio test for number of components is shown in Table 6.4. These point to three components for the data, correctly matching the three different species of iris that the data are known to contain. Group three corresponds exactly to *Iris setosa*, group two consists of 45 of the *Iris versicolor* plants and group one, all the *Iris virginica* plants plus the remaining five *Iris versicolor*.

A further example of the use of multivariate normal mixtures in cluster analysis is provided by the study reported in Powell, Clark and Bailey (1979).

Table 6.2 Data for an unfilled survey of the Corona Borealis region (Reproduced with permission from Roeder, 1990)

Velocity (km per second)		
9,172	9,350	9,483
9,558	9,775	10,227
10,406	16,084	16,170
18,419	18,552	18,600
18,927	19,052	19,070
19,330	19,343	19,349
19,440	19,473	19,529
19,541	19,547	19,663
19,846	19,856	19,863
19,914	19,918	19,973
19,989	20,166	20,175
20,179	20,196	20,215
20,221	20,415	20,629
20,795	20,821	20,846
20,875	20,986	21,137
21,492	21,701	21,814
21,921	21,960	22,185
22,209	22,242	22,249
22,314	22,374	22,495
22,746	22,747	22,888
22,914	23,206	23,241
23,263	23,484	23,538
23,542	23,666	23,706
23,711	24,129	24,285
24,289	24,366	24,717
24,990	25,633	26,960
26,995	32,065	32,789
34,279		

Four tests scores from the *Minnesota Test for the Differential Diagnosis of Aphasia* (see Schuell, 1965), were recorded for 86 patients referred to the Speech Therapy department of a hospital. The results from fitting a four component mixture with covariance matrices assumed to be equal are shown graphically in Figure 6.5. (The means shown are the mean error score expressed as a percentage of the maximum possible number of errors). The interpretation of the four groups was that they differed broadly in severity.

Several other examples of the application of multivariate normal mixtures are given in McLachlan and Basford (1988).

6.6 Mixtures for categorical data—latent class analysis

The mixture models considered in the previous section will, of course, not be suitable for data sets where the variables are categorical, since they assume

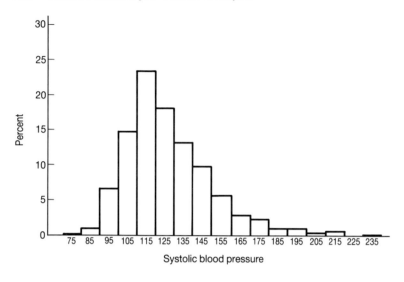

Systolic blood pressure

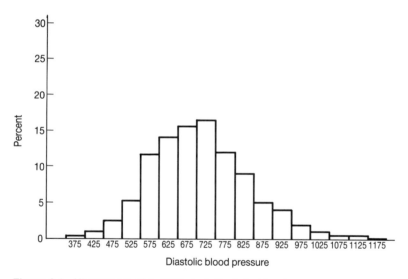

Diastolic blood pressure

Figure 6.4 Histograms of systolic and diastolic blood pressures.

that, within each group, the variables have a multivariate normal distribution. Clearly to provide suitable models for categorical data, the mixture assumed will need to involve other component densities. Most widely used are *multivariate Bernoulli* densities, which arise from assuming that, within each group, the categorical variables are independent of one another, the so called *condi-*

Table 6.3 Means and variances for the two derived normal distributions, whites and non-whites, all ages

	Mean	Variance	No. of subjects	Percent
		Whites		
Systolic				
Group 1	118·3	215	847	74
Group 2	147·6	694	301	26
Diastolic				
Group 1	65·7	102	821	72
Group 2	78·4	169	325	28
		Non-whites		
Systolic				
Group 1	116·1	159	136	65
Group 2	145·9	552	74	35
Diastolic				
Group 1	71·0	102	178	85
Group 2	94·9	54	30	15

Table 6.4 Likelihood ratio results from fitting multivariate normal mixtures to the iris data

Hypotheses	Log of likelihood ratio	Chi-square	df	p
2 against 1 cluster	165·56	318·98	28	< 0·001
3 against 2 clusters	34·17	65·60	28	< 0·001
4 against 3 clusters	13·51	25·86	28	0·06

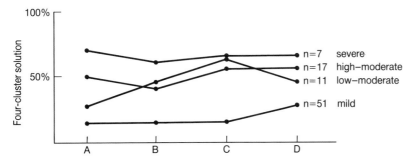

Figure 6.5 Four group solution for data on aphasic patients.

tional independence assumption. It is this approach which is the basis of *latent class analysis*, (see Lazarsfeld and Henry, 1968; Goodman, 1974). Here the appropriate model for binary variables is described.

Suppose that there are g groups in the data and that in group i, the vector $\boldsymbol{\theta}_i$ gives the probabilities of a 'one' response on each variable, that is

$$\Pr(x_{ij} = 1/\text{group } i) = \theta_{ij} \qquad (6.14)$$

where x_{ij} is the value taken by the jth variable in group i. From the conditional independence assumption it follows that the probability of an observed vector of scores, $\mathbf{x}$, in group i is given by

$$f(\mathbf{x}/\text{group } i) = \prod_{j=1}^{p} \theta_{ij}^{x_{ij}} (1 - \theta_{ij})^{1 - x_{ij}} \qquad (6.15)$$

If the proportions of each group in the population are $p_1, p_2 \cdots, p_g$ then the unconditional probability of the observation $\mathbf{x}$ is given by the mixture density

$$\Pr(\mathbf{x}) = \sum_{i=1}^{g} p_i \prod_{j=1}^{p} \theta_{ij}^{x_{ij}} (1 - \theta_{ij})^{1 - x_{ij}} \qquad (6.16)$$

Estimation of the parameters is again via maximum likelihood, and formation of groups by considering the estimated posterior probabilities—see McLachlan and Basford (1988) for details.

6.6.1   Applications of latent class analyis

As part of their detailed statistical modelling of an extensive body of educational research data on teaching, Aitkin, Anderson and Hinde (1981) fitted latent class models to observations on 38 binary variables describing teaching behaviour observations made on 468 teachers. The parameter estimates obtained by maximum likelihood via the EM algorithm for the two and three class models are shown in Table 6.5. For the two class model, the response probabilities marked * show large differences between the classes, indicating systematic differences in behaviour on these items for teachers in the two groups. For the three class model the response probabilities for Classes 1 and 2 are very close to those for the corresponding classes in the two class model (though in most cases more widely separated), and the response probabilities for Class 3 are mostly between those for Classes 1 and 2. Thus Class 3 is to some extent intermediate between Classes 1 and 2. Aitkin, Anderson and Hinde interpreted the groups they found in terms of 'formal' and 'informal' teaching styles.

A further interesting application of latent class analysis is reported in Pickering and Forbes (1984), where the method is used to study how to allocate neonatal resources throughout Scotland. By classifying individuals into groups with similar medical characteristics the resources required by each area can be estimated from the prevalance of each type. The data for analysis consisted of 11 categorical variables observed on 45,426 cases. The variables contained clinical and diagnostic information extracted from the Scottish Neonatal Discharge Record. Two, three and four class models were fitted using maximum

Table 6.5 Two- and three-latent class parameter estimates $(100 \times \hat{\theta}_{ij})$ for teacher data (Reproduced with permission from Aitkin *et al*, 1981)

		Item	Two-class model		Three-class model		
			Class 1	Class 2	Class 1	Class 2	Class 3
1 (1)		Pupils have choice in where to sit	22	43	20	44	33
2		Pupils sit in groups of three or more	60	87*	54	88	79
3 (2)		Pupils allocated to seating by ability	36	23	36	22	30
4		Pupils stay in same seats for most of day	91	63*	91	52	89
5 (3)		Pupils not allowed freedom of movement in classroom	97	54*	100	53	74
6		Pupils not allowed to talk freely	89	48*	94	50	61
7		Pupils expected to ask permission to leave room	97	76*	96	69	95
8 (4)		Pupils expected to be quiet	82	42*	92	39	56
9		Monitors appointed for special jobs	85	67	90	70	69
10 (5)		Pupils taken out of school regularly	32	60	33	70	35
11		Timetable used for organizing work	90	66*	95	62	77-
12		Use own materials rather than textbooks	19	49	20	56	26
13		Pupils expected to know tables by heart	92	76	97	80	75†
14		Pupils asked to find own reference materials	29	37	28	39	34
15 (6)		Pupils given homework regularly	35	22	45	29	12†
16 (i)	(7)	Teacher talks to whole class	71	44	73	37	62
(ii)	(8)	Pupils work in groups on teacher tasks	29	42	24	45	38
(iii)	(9)	Pupils work in groups on work of own choice	15	46*	13	59	20
(iv)	(10)	Pupils work individually on teacher tasks	55	37	57	32	50
(v)	(11)	Pupils work individually on work of own choice	28	50	29	60	26†
17		Explore concepts in number work	18	55*	14	62	34
18		Encourage fluency in written English even if inaccurate	87	94	87	95	90
19 (12)		Pupils' work marked or graded	43	14*	50	16	20
20		Spelling and grammatical errors corrected	84	68	86	64	78
21 (13)		Stars given to pupils who produce best work	57	29	65	30	34
22 (14)		Arithmetic tests given at least once a week	59	38	68	43	35†
23 (15)		Spelling tests given at least once a week	73	51	83	56	46†
24		End of term tests given	66	44	75	48	42†
25		Many pupils who create discipline problems	09	09	07	01	18†
26		Verbal reproof sufficient	97	95	98	99	91†
27 (i)		Discipline—extra work given	70	53	69	49	67
(ii)	(16)	Smack	65	42	64	33	63
(iii)		Withdrawal of privileges	86	77	85	74	85
(iv)		Send to head teacher	24	17	21	13	28†
(v)	(17)	Send out of room	19	15	15	08	27†
28 (i)	(18)	Emphasis on separate subject teaching	85	50*	87	43	73
(ii)		Emphasis on aesthetic subject teaching	55	63	53	61	63†
(iii)	(19)	Emphasis on integrated subject teaching	22	65*	21	75	33
p.		Estimated proportion of teachers in each class	0·538	0·462	0·366	0·312	0·322

* An item with large differences in response probability between Classes 1 and 2.
† An item in which Class 3 is extreme.

Table 6.6   Classification of Scottish infants. Parameter estimates for one to four class models using 1980 data and complete cases only (Reproduced with permission from Pickering & Forbes, 1984)

Vari-able	No of levels	Levels‡	1 class I	2 classes IIa	IIb	3 classes IIIa	IIIb	IIIc	4 classes IVa	IVb	IVc	IVd
1	4	2001–2500 g	0·05	0·01	0·48	0·00	0·18	0·79	0·00	0·20	0·78	0·03
		1501–2000 g	0·01	0·00	0·15	0·00	0·26	0·09	0·00	0·32	0·09	0·00
		≤ 1500 g	0·01	0·00	0·08	0·00	0·21	0·01	0·00	0·25	0·01	0·00
2	2	< 10th centile	0·10	0·07	0·43	0·07	0·19	0·62	0·07	0·21	0·62	0·10
3	2	< 7	0·02	0·01	0·12	0·01	0·26	0·01	0·00	0·21	0·01	0·32
4	3	Intermediate*	0·09	0·08	0·19	0·08	0·26	0·13	0·07	0·25	0·13	0·31
		By intubation	0·03	0·02	0·17	0·02	0·33	0·04	0·01	0·29	0·00	0·52
5	2	Present	0·01	0·00	0·10	0·00	0·25	0·00	0·00	0·29	0·00	0·01
6	2	Present	0·01	0·00	0·06	0·00	0·17	0·00	0·00	0·20	0·00	0·00
7	2	Present	0·30	0·28	0·58	0·28	0·67	0·49	0·28	0·71	0·49	0·32
8	2	Present	0·00	0·00	0·03	0·00	0·07	0·00	0·00	0·07	0·00	0·01
9	2	Present	0·03	0·01	0·30	0·01	0·60	0·10	0·01	0·67	0·10	0·05
10	2	Present	0·00	0·00	0·05	0·00	0·13	0·00	0·00	0·15	0·00	0·00
11	3	4–10 days	0·80	0·84	0·34	0·83	0·09	0·53	0·83	0·04	0·50	0·84
		> 11 days	0·08	0·03	0·61	0·03	0·79	0·45	0·03	0·82	0·46	0·14
Frequency of class			1·00	0·92	0·08	0·92	0·03	0·05	0·89	0·03	0·05	0·04

* Mask + intermittent positive pressure ventilation, Drugs only, Other.
† > 86 μmol/litre bilirubin
‡ Parameters for all levels of a variable sum to 1, the first level is omitted without loss of information

Key:

1 Birthweight	4 Resuscitation	7 Jaundice†	10 Dead at discharge
2 Birthweight for gestation age	5 Assisted ventilation after 30 min	8 Convulsions	11 Age at discharge
3 Apgar at 5 min	6 Recurrent apnoea	9 In tube feeding	

likelihood methods. The results are shown in Table 6.6. Of the solutions shown, that with four groups was considered by the investigators as the most interesting and relevant. One class was identified as healthy infants, two others were associated with moderately ill infants requiring various types of special neonatal care and the fourth class contained severely ill, very low birthweight infants. How the classification was finally used is discussed in detail in the original paper.

6.7   Mixture models for mixed mode data

Multivariate normal mixtures are applicable to data containing continuous variables, latent class models to data with categorical variables. In practice, of course, many data sets will contain variables of both types, and Everitt (1988) suggests how the mixture approach can be applied to such *mixed mode* data.

Suppose the data consists of p continuous variables $\mathbf{x}' = [x_1, x_2, \ldots, x_p]$

and q categorical variables, $\mathbf{z}' = [z_1, z_2, \ldots, z_q]$, with z_i having c_i categories. It is assumed that each categorical variable, z_i arises from an unobservable continuous random variable, y_i, by applying a series of thresholds, i.e.

$$z_i = k \quad \text{if} \quad \alpha_{i,k-1} \leqslant y_i < \alpha_{i,k} \tag{6.17}$$

for $k = 1, 2, \ldots, c_i$ and $\alpha_{i,0} = -\infty$, $\alpha_{i,c_i} = \infty$, $i = 1, 2 \ldots, q$. The threshold values α_{ij}, $i = 1, 2, \ldots, q$, $k = 1, 2, \ldots, c_i - 1$ are considered as unknown parameters to be estimated from the data.

The density function of $\mathbf{x}, \mathbf{y} = [y_1, y_2, \ldots, y_q]$ is assumed to be the usual multivariate normal mixture, with assumed common covariance matrix. The density function of the observed variables $\mathbf{x}, \mathbf{z}$ is then given by

$$h(\mathbf{x}, \mathbf{z}) = \sum_{i=1}^{g} p_i \int_{a_1}^{b_1} \int_{a_2}^{b_2} \cdots \int_{a_q}^{b_q} \alpha(\mathbf{y}, \boldsymbol{\mu}_i, \boldsymbol{\Sigma}) dy_1 dy_2 \ldots dy_q \tag{6.18}$$

where the limits of integration correspond to the threshold values appropriate to the particular values in the vector $\mathbf{z}$.

Maximum likelihood can be used to estimate the parameters of the model, although the presence of the multidimensional integral in (6.18) makes the procedures computationally impractable if the number of categorical variables is greater than four. Examples of how this method performs in comparison with other approaches to the clustering of mixed mode data are given in Everitt and Merette (1989).

6.8 Summary

The mixture approach to cluster analysis provides a firmer statistical basis than methods discussed in previous chapters, and involves no decision over what particular similarity or distance measure is appropriate for a data set. Nevertheless mixture models have their own set of assumptions, for example normality, conditional independence, which may not be realistic in all applications.

7

Other Clustering Techniques

7.1 Introduction

The methods described in the preceeding chapters form the major part of the body of work on cluster analysis. Nevertheless there remains a substantial number of other methods which have been developed but do not fall clearly into any of the previous categories. In this chapter an attempt is made to describe a number of these techniques, although a comprehensive review is impossible because of the vastness of the literature involved.

7.2 Density search clustering techniques

If individuals are depicted as points in a metric space, a natural concept of clustering (see Chapter 1) suggests that there should be parts of the space in which the points are very dense, separated by parts of low density. Several methods of cluster analysis have been developed which search for regions of high density in the data, each such region generally being taken to signify a different group. (The multivariate normal mixture approach described in Chapter 6 might be seen as a formal way of using this concept). A number of these techniques have their origins in single linkage clustering (see Chapter 4) but attempt to overcome chaining, one of the main problems with the method. As mentioned in Chapter 4 this refers to the tendency to incorporate individuals into existing clusters rather than to initiate new ones.

7.2.1 The taxmap method

Carmichael *et al.* (1968) and Carmichael and Sneath (1969), describe a clustering method which attempts to imitate the procedure used by the human observer for detecting clusters in two or three dimensions, namely to compare relative distances between points and to search for continuous, relatively densely populated regions of the space, surrounded by continuous relatively empty spaces. Clusters are formed initially in a way similar to that for the single linkage method, but criteria are adopted for judging when additions to clusters should be stopped. One such criterion is to terminate additions if the prospective point is 'much' further away than was the last point admitted, as indicated by a discontinuity in some measure of closeness. For the latter, the drop in the average similarity on addition of an individual to the cluster, from the new average, is used. This measure has been found to decrease smoothly up to a discontinuity, whereas the drop in the average similarity itself varies widely.

To illustrate the method consider the following matrix of similarities between five individuals:

$$
\begin{array}{c}
 \\
 \\
S = \\
 \\

\end{array}
\begin{array}{c}
1 \\
2 \\
3 \\
4 \\
5
\end{array}
\left(
\begin{array}{ccccc}
1 & 2 & 3 & 4 & 5 \\
1.0 & & & & \\
0.7 & 1.0 & & & \\
0.9 & 0.8 & 1.0 & & \\
0.4 & 0.5 & 0.4 & 1.0 & \\
0.3 & 0.4 & 0.2 & 0.7 & 1.0
\end{array}
\right)
$$

The two most similar individuals are used to initiate the cluster. From **S** these are found to be individuals 1 and 3 with similarity 0.9. The next individual considered for admission to the cluster is the one most similar to an individual already in the cluster (cf. single linkage). This leads to consideration of individual number 2 whose similarity with individual 3 is 0.8. The average similarity between the three individuals is now calculated to give

Cluster members	Candidate individual
1, 3	2
Similarity 0.9	

The average similarity between individuals 1, 2 and 3 is

$$\frac{1}{3}(0.9 + 0.7 + 0.8) = 0.8$$

Therefore the drop in similarity is $0.9 - 0.8 = 0.1$, and the measure of discontinuity is $(0.8 - 0.1) = 0.7$. Low values of this measure indicate that the candidate point should not be added to the cluster. Suppose a low value is taken to be one less than 0.5, then individual 2 would be added to the cluster and a further individual would be considered for admission. The candidate individual is now individual 4.

Cluster members	Candidate individual
1, 2, 3	4
Similarity 0.8	

The average similarity between individuals 1, 2, 3 and 4 is

$$\frac{1}{6}(0.9 + 0.7 + 0.4 + 0.8 + 0.5 + 0.4) = 0.6$$

Therefore the drop in similarity is 0.2, and the measure of discontinuity is 0.4. Consequently the individual is not admitted to the cluster but instead initiates a new cluster.

Various other criteria are used to prevent admissions of points relatively near the centroid of an elongated cluster but still rather far from any point in the cluster.

7.2.2   Mode analysis

Mode analysis (Wishart, 1969a), is a derivative of single linkage clustering which searches for natural sub-groupings of the data by seeking disjoint density surfaces in the sample distribution. The search is made by considering a sphere of some radius, R, surrounding each point and counting the number of points falling in the sphere. Individuals are then labelled as *dense* or *non-dense* depending on whether their spheres contain more or less points than the value of the *linkage* parameter, K, which is preset at a value dependent on the number of individuals in the data set. (Some possible values of K for various values of n are suggested in Wishart, 1978).

The parameter R is gradually increased and so more individuals became 'dense'. Four courses of action are possible with the introduction of each new dense point.

(i)   The new point is separated from all other dense points by a distance which exceeds R. When this happens the point initiates a new cluster nucleus and the number of clusters is increased by one.

(ii)   The new point is within distance R of one or more dense points which belong to only one cluster nucleus. In this case, the new point is added to the existing cluster.

(iii) The new point is within distance R of dense points belonging to two or more clusters. If this happens the clusters concerned are combined.

(iv)   At each 'introduction' cycle the smallest distance, D, between dense points belonging to different clusters is found, and compared with a threshold value calculated from the average of the $2K$ smallest distance coefficients for each individual. If D is less than this threshold value then the two clusters are combined. Sometimes only one cluster is produced (indicating a lack of cluster structure in the data), but usually the analysis reaches a point at which a maximum number of clusters is isolated. It is usually this solution which is taken as the most significant.

A difficulty with mode analysis is its failure to identify both large and small clusters simultaneously. A small radius R may distinguish two large disjoint modes without finding a third smaller, but distinct mode, because each of its individuals fails to qualify as a dense neighbourhood. Alternatively if a larger R is specified the small cluster might be found, but the two large clusters could possibly be merged. Such potential difficulties led Wishart (1973) to suggest an improved mode-seeking cluster method in which the spherical neighbourhoods of two growing clusters may intersect at some large value of R, to the extent

Table 7.1   Results from applying mode analysis to the
bee distance matrix, Table 4.1

Stage	
1	Observation 3 initiates new cluster centre
2	Observation 5 joins observation 3
3	Observation 4 joins [3, 5]
4	Observation 6 joins [3, 5, 4]
5	Observation 7 joins [3, 5, 4, 6]
6	Observation 10 initiates new cluster centre
7	Observation 9 joins observation 10
8	Observation 2 joins [3, 5, 4, 6, 7]
9	Observation 8 joins [3, 5, 4, 6, 7, 2]
10	Observation 1 joins [3, 5, 4, 6, 7, 2, 8]
11	Observation 11 joins [10, 9]

Cluster 1: 3, 5, 4, 6, 7, 2, 8, 1
Cluster 2: 10, 9, 11

that they would have been fused in the original version of mode analysis, but now the fusion level is merely noted, and the clusters are not united.

As an illustration of the use of mode analysis it was applied to the distance matrix for eleven forms of the bee *Hoplites producta* given previously in Chapter 4. The results are shown in Table 7.1.

7.2.3   Nearest neighbour clustering procedures

Wong (1982) and Wong and Lane (1983) describe a hierarchical clustering method which is similar in some respects to the method of mode analysis discussed in the previous section. The method is designed to detect what Hartigan (1975) defines as *high-density clusters*, these being maximal connected sets of the form

$$\{x|f(x) \geqslant f^*\} \tag{7.1}$$

where f is the population density of the observations, and f^* is some threshold value. Wong and Lane (1983) estimate the density at a point x by $f_n(x)$ given by

$$f_n(x) = k/(nV_k(x)) \tag{7.2}$$

where $V_k(x)$ is the volume of the smallest sphere centred at x containing k sample observations. A distance matrix arises from these density estimates according to the following two definitions

(i) Definition 1.   Two observations x_i and x_j are said to be *neighbours* if $d^*(x_i, x_j) \leqslant d_k(x_i)$ or $d_k(x_j)$ where d^* is the Euclidean metric and $d_k(x_i)$ is the kth nearest neighbour distance to point x_i.

(ii) Definition 2.   The distance between the observations x_i and x_j is

$$(1/2)[1/f_n(x_i) + 1/f_n(x_j)] \tag{7.3}$$

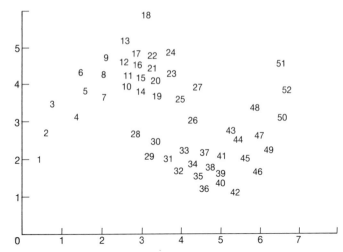

Figure 7.1   Bivariate data containing crescentric clusters. (Reproduced with permission from Wong and Lane, 1983.)

$$= \frac{n}{2k}\left[V_k(x_i) + V_k(x_j)\right] \text{ if } x_i \text{ and } x_j \text{ are neighbours}$$
$$= \infty, \text{ otherwise}$$

The single linkage clustering algorithm is then applied to this distance matrix to obtain the dendrogram of sample high-density clusters.

The value of k controls the amount by which the data are 'smoothed' to give the density estimate on which the clustering procedure is based. There appears to be no unique recommendation concerning the choice of k, although Wong and Schaak (1982) derive empirical evidence for a rule of thumb of the form

$$k = 2\log_2 n \qquad (7.4)$$

Since the hierarchical clusterings obtained for different values of k can be very different, Wong and Lane (1983) suggest that several values around that given in equation (7.4) should be tried.

To illustrate the operation of their proposed method, Wong and Lane (1983) apply it to the data shown in Figure 7.1. The dendrogram giving the hierarchical clustering obtained by the kth nearest neighbour method with $k = 5$ appears in Figure 7.2. Two disjoint modal regions, corresponding to the crescentric clusters in Figure 7.1 can be identified in this dendrogram.

Ling (1972) suggests another nearest neighbour type clustering procedure, and Wong and Schaak (1982) propose a method for indicating the number of clusters when using the method outlined above.

Other density search type clustering methods are described in Gitman and Levine (1970), Cattell and Coulter (1966) and Katz and Rohlf (1973).

Obs. number

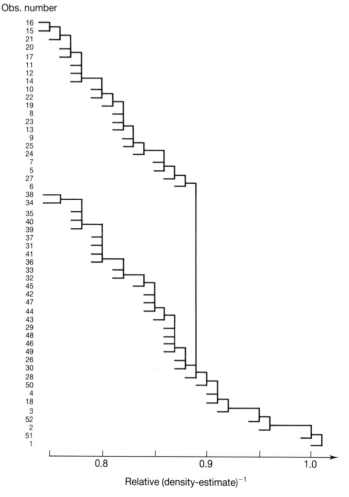

Figure 7.2   Dendrogram obtained by applying Wong and Lane clustering to the data shown in Figure 7.1. (Reproduced with permission from Wong and Lane, 1983.)

7.3   Clumping techniques

Most clustering methods lead to distinct or disjoint clusters, but there are a number of techniques available which allow overlapping clusters. Such methods are often referred to as *clumping* techniques, a term first introduced by Jones, Needham and their fellow workers at the Cambridge Language Research Unit.

Clumping techniques often begin with the calculation of a similarity matrix, followed by the division of the data into two groups by minimizing what is known as a *cohesion function*. Needham (1967) considered a symmetric cohesion function, $G_1(A)$ given by

$$G_1(A) = \frac{S_{AB}}{S_{AA}S_{BB}} \tag{7.5}$$

and Parker-Rhodes and Jackson (1969) suggest a modification, $G_2(A)$ given by

$$G_2(A) = \frac{S_{AB}}{S_{AA}} \left[\frac{n_A(n_A - 1)}{S_{AA}} - \frac{S_{AA}}{bn_A(n_A - 1)} \right] \tag{7.6}$$

where A and B refer to the two groups into which the data are divided, A being their putative 'clump'. S_{XY} is the sum of the similarities between members of groups X and Y, that is

$$S_{XY} = \sum_{i \in x} \sum_{j \in Y} s_{ij}$$

where s_{ij} is an inter-individual similarity coefficient, n_A, is the number of individuals in group A, and b is an arbitrary parameter, which allows the investigator some control over the size of the 'clumps' and the amount of overlap.

Algorithms to minimize these functions proceed by successive reallocations of single individuals from an initial randomly chosen cluster centre (Jones and Jackson, 1967; Bonner, 1964). By iterating from different starting points many divisions into two groups may be found. In each case the members of the smaller group are noted and constitute a class to be set aside for further examination. The cohesion function $G_1(A)$ is designed to find good partitions of the set of individuals, whilst $G_2(A)$ allows the internal similarities of A and the separation of A from B to be adjusted relative to each other by use of the parameter b.

Sibson (1970) shows that axioms of stability, optimal cluster preservation and invariance under relabelling or any monotonic transformation of the proximity matrix, lead uniquely to a system first described by Jardine and Sibson (1968) which generates a series of overlapping clusters. This method consists of representing each individual by a node on a graph and connecting all pairs of nodes which correspond to individuals having a similarity value above some specified threshold, H. Next a search is made for the largest set of individuals for which all pairs of nodes are connected (these are known as *maximal complete subgraphs*). Then all pairs of maximal complete subgraphs which intersect in at least a particular number of nodes, say K, are further connected. When no more connections can be found the solution for the particular values of H and K considered has been obtained. When $K = 1$ no overlaps occur and the procedure reduces to single linkage clustering. In general the clusters found by this method can overlap to the extent of having $K - 1$ points in common.

The procedure is illustrated in Figure 7.3 for the similarity matrix shown in Table 7.2. An algorithm given by Jardine and Sibson (1968) for implementing

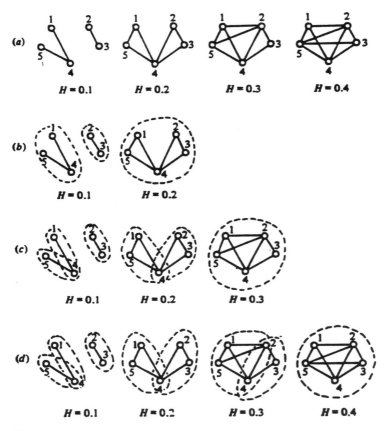

Figure 7.3 Results from applying Jardine and Sibson's clustering method to the similarity matrix in Table 7.2.

Table 7.2   Hypothetical similarity matrix

2	0·3			
3	0·4	0·1		
4	0·1	0·2	0·2	
5	0·2	0·3	0·4	0·1

the method has been considerably improved by Cole and Wishart (1970). Further discussion of Jardine and Sibson's clustering procedure is given in Rohlf (1974, 1975) and an example of its use in practice is described in Morgan (1973).

7.4   Constrained classification

In some problems in classification, there are external constraints on the individuals which are to be classified. Gordon (1973), for example, describes a situation which occurs in palaeoecology, where the individuals to be classified are samples down a vertical bore and it is required that the stratigraphical information is preserved in any classification. If the samples are labelled $1, 2, \ldots, n$ according to level, and g groups are sought, then $g - 1$ 'markers' are needed in some of the $n - 1$ gaps between pairs of neighouring samples. The groups would be as follows

$$1 \text{ to } n_1, |n_1 + 1 \text{ to } n_2| \ldots |n_{g-1} + 1 \text{ to } n|$$

The number of ways in which $g - 1$ markers can be located, and hence the number of possible partitions of the n individuals into g groups of contiguous individuals is

$$\frac{(n - 1)!}{(g - 1)!(n - g)!} \tag{7.7}$$

which represents a considerable reduction in possible partitions from the unconstrained case (see Chapter 5).

The markers are placed so as to ensure that some measure of within group variation is small. Any of the measures described in Chapter 5 might be used, but generally the within group sum of squares criterion is employed. The optimal partition into g groups for a range of values of g is sought. The solutions obtained need not be hierarchically nested; the optimal three groups, for example, need not be obtainable from the optimal two groups by subdivision of one of these groups.

Gordon (1980) suggests two approaches to finding the optimal partition. The first uses a divisive algorithm which begins by finding a single marker leading to minimum within group sum of squares. Each of the two groups is then optimally divided, choosing the division which leads to maximum decrease in the sum of squares. The algorithm continues by successive division of existing groups.

The second procedure described by Gordon (1980) involves a dynamic programming algorithm. The details are as follows:

Let $s(i, j)$ denote the within group sum of squares of objects i to j inclusive, and let $t(g, k)$ denote the total within group sum of squares when objects 1 to k are optimally divided into g groups. It is $t(g, n)$ for $2 \leqslant g \leqslant n$, together with the corresponding markers that is required. The solution is built up recursively, evaluating $\{t(g, k), k = g, g + 1, \ldots, n; g = 1, 2, \ldots, n\}$ by use of the following formulae:

$$t(1, k) = s(1, k) \quad (1 \leqslant k \leqslant n) \tag{7.8}$$

$$t(g, k) = \min_{\substack{g-1 \leqslant i \\ \leqslant k-1}} [t(g - 1, i) + s(i + 1, k)] \quad (g \leqslant k \leqslant n; \ 2 \leqslant g \leqslant n) \tag{7.9}$$

Equation (7.9) involves dividing the first k objects into two classes, the first class containing $g - 1$ groups and the second class containing one group of objects. This is equivalent to placing the last marker at position i, and the algorithm finds the optimal value of i. The complete set of $g - 1$ markers is obtained using a trace-back procedure.

Each of these approaches was applied to the data shown in Table 7.3, (kindly supplied by Professor I.T. Jolliffe). The observations here are the separate 10km times for the 80 competitors in a 100km road race. The data are ordered in terms of finishing times, and it is finishing position which determines the constraint. The results of applying the divisive and the dynamic programming algorithms to these data are shown in Table 7.4. (These analyses were generously carried out by Dr A. Gordon). For both methods the optimal division into two groups occurs at the mid-way finishing position, so that group one consists of finishers 1–40 and group two finishers 41–80. It is this division which produces the greatest decrease in the sum of squares criterion. The two methods give different solutions for other values of g, although both indicate that groups containing finishers 1–5 and finishers 74–80, are reasonably stable.

Other approaches to the constrained classification problem are described in Ferligoj and Batagelj (1982) and Lechevallier (1980).

7.5   Simultaneous clustering of individuals and variables

Most clustering procedures are designed to construct an optimal partition of the individuals in a data set or to build up a hierarchy of classes. 'Clustering' of variables is usually regarded as the province of a technique such as factor analysis, although, as mentioned in Chapter 1 there is no reason that particular clustering techniques should not be used for this purpose. There are situations however where a classification of individuals and of variables may be needed *simultaneously*. An example given by Bock (1980) is a country which has n districts and its residents originate from m different nations with x_{ij} the proportion of inhabitants from nation j in the district i. In a sociological study it may be useful to search, simultaneously, for a partition of districts and another partition of nationalities such that the demographic structure shall be evident and the interaction between the classes of regions and the national classes will result.

Hartigan (1972) also presents an example for which separate clustering of individuals and variables fails to reveal the most interesting structure in a data set. The data consist of the Republican votes for president of the USA in a number of Southern States for a number of years in the 20th century. The data are shown in Table 7.5, along with the Euclidean distance matrices for years and for states. Corresponding complete linkage dendrograms are shown in Figure 7.4. Years 32, 40 and 36 form one group, 60 and 68 another and 64 a third. For the states there appear to be two main clusters, SC, LA, MI and KY, MD, MO. But this information does not reveal the main happening in the data, namely the unusual interaction between SC, LA, MI and 60, 68, 64.

The approach taken by Bock (1980) to this problem is to search for a partition $R = (I_1, I_2, \ldots, I_{gr})$ of the individuals and for a partition $C = (V_1, V_2, \ldots, V_{gc})$ of the variables such that the pair (R, C) reflects optimally the 'structure' of the data set. The term optimal is made precise by introducing a clustering criterion which measures the 'deviation' between a certain partition

Table 7.3 10 km times in 100 km road race

Position	\ 10 km time (minutes)									
	1st	2nd	3rd	4th	5th	6th	7th	8th	9th	10th
1	37·0	37·8	36·6	39·6	41·0	41·0	41·3	45·7	45·1	43·1
2	39·5	42·2	40·0	42·3	40·6	40·8	42·0	43·7	41·0	43·9
3	37·1	38·0	37·7	42·2	41·6	43·5	48·7	49·7	44·8	47·0
4	37·0	37·8	36·6	39·6	41·0	44·8	44·5	49·4	44·6	47·7
5	42·2	44·5	41·9	43·4	43·0	47·2	49·1	49·9	46·8	52·3
6	43·0	44·6	41·2	42·1	42·5	46·8	47·5	55·8	56·6	58·6
7	43·2	44·4	41·0	43·4	43·0	47·2	52·4	57·3	54·4	53·5
8	43·2	46·7	44·8	47·5	47·4	47·7	49·9	52·1	50·7	50·0
9	38·5	41·4	40·1	43·2	43·2	51·5	56·7	71·5	56·2	48·2
10	42·5	43·1	40·6	44·5	45·4	52·3	59·7	59·3	55·0	49·6
11	38·0	40·1	39·1	43·8	46·6	51·9	59·2	63·5	57·6	58·4
12	46·0	50·4	46·8	47·4	44·1	43·4	46·3	55·0	64·9	56·2
13	44·8	46·0	43·1	46·5	46·3	49·0	52·5	58·4	60·9	55·2
14	44·8	46·0	43·1	46·5	46·3	49·0	52·5	58·4	60·9	55·2
15	47·0	49·4	46·8	48·6	47·8	50·8	50·3	54·0	54·4	53·6
16	45·0	46·7	45·3	49·9	47·8	51·2	54·1	58·7	53·3	50·7
17	45·0	46·7	43·8	48·0	47·2	47·5	51·7	57·3	60·4	55·6
18	43·1	44·5	41·0	42·5	40·6	42·8	46·5	73·2	70·8	63·4
19	45·2	46·9	45·5	48·8	50·1	51·2	56·4	55·2	56·6	53·5
20	43·0	46·1	44·7	47·4	47·1	46·8	54·6	60·4	68·0	51·6
21	38·3	41·6	39·6	40·7	41·6	41·6	47·2	62·4	82·7	77·1
22	45·0	47·1	45·3	49·1	46·8	47·4	50·3	55·1	66·4	64·6
23	43·2	46·1	45·2	48·4	49·9	49·6	52·7	58·1	62·8	62·6
24	41·2	44·6	43·8	48·4	48·8	53·4	58·9	68·6	59·1	53·4
25	49·2	48·8	48·7	51·8	48·2	52·8	50·2	58·0	58·7	57·5
26	48·0	52·9	49·6	50·1	48·1	48·1	49·1	54·6	62·7	64·0
27	46·0	49·9	47·7	50·4	52·9	51·4	55·6	57·8	59·7	55·8
28	46·1	46·0	42·2	44·4	46·0	49·0	53·3	66·7	72·9	67·6
29	48·0	52·9	49·6	50·1	48·4	50·0	58·5	62·9	60·1	60·1
30	45·1	49·7	46·5	46·5	49·3	58·8	58·7	64·7	64·0	63·4
31	49·2	54·5	51·3	56·1	53·9	53·2	53·4	58·8	62·4	59·4
32	47·0	49·4	46·8	49·7	50·3	55·5	59·8	67·1	64·2	70·4
33	48·2	54·1	51·2	53·5	54·8	55·7	55·2	65·7	62·3	62·3
34	46·5	50·8	48·0	51·4	50·0	58·6	61·6	61·5	61·9	75·4
35	47·3	51·2	49·5	52·6	57·9	58·6	66·4	70·6	56·4	55·6
36	48·2	53·9	50·9	54·0	52·4	59·3	77·5	60·6	55·8	61·4
37	43·3	45·2	42·7	44·9	47·3	52·9	69·3	92·2	57·3	79·1
38	52·0	53·0	50·0	51·6	55·4	56·3	56·7	68·4	66·9	65·4
39	49·2	54·5	50·8	53·6	53·4	56·0	62·3	65·8	66·1	65·6
40	49·3	52·8	51·1	53·8	52·4	59·3	63·2	73·7	62·3	62·3
41	47·2	51·3	49·5	52·6	51·6	60·1	64·4	66·5	66·6	76·9
42	49·2	48·8	49·2	54·2	60·8	60·4	64·0	69·9	66·1	65·1
43	45·2	50·2	48·7	53·6	53·5	60·3	59·2	71·4	75·9	71·8
44	45·3	51·0	46·9	50·0	51·0	59·7	78·2	68·9	69·7	72·8
45	49·2	48·3	46·2	51·6	51·9	61·1	71·8	74·6	70·3	69·9
46	49·2	48·7	48·8	52·6	57·9	65·3	71·7	64·1	70·7	68·2
47	46·0	49·2	47·5	51·5	54·1	61·4	66·5	76·5	77·0	68·6

Table 7.3   *continued*

Position	1st	2nd	3rd	4th	5th	6th	7th	8th	9th	10th
					10 km time					
48	46·5	51·1	49·9	56·1	53·6	58·2	66·3	76·2	70·6	76·3
49	51·6	54·0	52·1	55·1	57·4	61·0	63·4	70·2	73·4	67·9
50	47·2	50·0	48·1	50·8	55·3	62·6	70·5	76·1	72·7	72·8
51	45·0	50·3	48·8	53·6	54·4	58·9	67·6	77·7	79·9	81·1
52	48·0	52·9	49·6	50·1	53·5	65·6	72·8	74·1	72·6	78·1
53	53·2	55·1	55·0	59·3	59·4	63·2	66·1	66·7	73·7	68·3
54	62·5	67·5	73·1	68·2	47·1	51·9	58·3	68·5	64·4	62·1
55	49·2	48·8	53·4	56·1	59·8	65·2	72·8	71·4	79·8	70·5
56	49·2	48·6	47·5	51·8	57·7	·63·5	63·5	69·5	92·6	83·4
57	51·6	53·7	49·2	58·3	56·4	65·3	74·8	75·4	75·8	69·2
58	58·7	62·7	56·3	58·6	66·3	62·9	67·4	71·4	69·6	60·8
59	49·2	53·3	53·7	54·8	59·3	73·9	70·8	86·3	61·8	78·7
60	59·0	64·6	61·4	64·0	60·2	64·0	66·2	69·5	69·3	64·9
61	50·1	53·9	52·7	59·8	58·2	71·4	72·3	78·4	77·5	74·9
62	55·0	58·5	59·4	63·4	57·0	66·4	67·7	68·7	75·9	77·7
63	47·2	52·1	51·7	61·0	73·2	74·5	69·2	76·5	75·6	70·9
64	51·7	54·0	53·0	55·6	56·0	62·9	76·2	81·1	85·5	87·8
65	50·0	47·3	44·1	51·7	62·8	75·3	78·1	81·2	85·5	87·8
66	56·2	59·7	55·6	58·2	64·4	76·1	68·4	75·3	84·3	70·5
67	56·2	59·7	55·6	58·2	64·4	76·1	68·4	75·3	84·8	70·5
68	46·5	51·7	52·3	61·7	66·8	68·1	76·9	74·9	83·7	96·1
69	56·2	60·0	60·4	67·7	64·7	73·1	68·7	72·1	70·3	86·9
70	49·2	53·0	52·5	55·5	57·1	77·7	86·6	71·2	82·5	104·6
71	51·6	54·2	58·7	59·8	65·4	76·2	73·1	93·0	83·7	74·3
72	58·1	62·0	60·2	63·7	65·7	78·8	69·4	81·7	79·2	81·9
73	48·2	54·0	52·5	55·5	60·1	73·9	71·8	86·8	91·9	110·0
74	55·0	60·9	55·0	63·5	68·8	84·6	64·8	95·0	81·8	75·4
75	56·2	60·4	63·1	65·0	71·7	78·8	77·0	89·0	83·5	61·0
76	48·0	52·9	52·8	70·5	77·1	85·3	76·9	93·9	88·4	68·2
77	46·5	51·0	63·6	66·7	75·0	81·0	76·0	95·4	80·9	79·3
78	46·5	51·0	63·6	66·7	75·0	81·0	76·0	95·4	80·9	79·3
79	52·2	55·5	55·9	70·6	77·7	86·6	71·6	86·9	87·8	71·2
80	50·5	55·4	64·1	66·3	75·6	86·6	71·6	87·3	89·2	73·4

pair and the given data matrix, **X**. In practice a variety of clustering criteria are possible and Bock (1980) considers three. They all rely on the assumption that a given pair of classes, I_k, V_l may be characterized by some unknown point u_{kl} such that all data points x_{ij} with $i \in I_k$ and $j \in V_l$ lie in the neighbourhood of u_{kl}. For full details of the method readers are referred to the original paper.

Table 7.4   Results of constrained clustering on data in Table 7.3

(i) Unweighted least squares analysis

No. of groups	Percent of total				Markers				
2	47·08	40							
3	37·36	40	62						
4	30·83	17	40	62					
5	27·30	17	40	62	73				
6	24·93	17	28	40	62	73			
7	22·55	5	17	28	40	62	73		
8	21·08	5	17	28	40	52	62	73	
9	19·95	5	17	28	40	52	62	72	73
10	18·95	5	17	28	40	52	54	62	72	73

(ii) Results of dynamic algorithm

No. of groups	Percent of total				Markers				
2	47·08	40							
3	34·01	31	62						
4	29·57	5	31	62					
5	25·84	5	31	60	73				
6	23·51	5	28	42	63	73			
7	22·04	5	28	40	52	63	73		
8	20·71	5	28	40	53	54	63	73	
9	19·66	5	17	31	42	53	54	63	73
10	18·50	5	20	21	31	42	53	54	63	73

7.6   Summary

In this chapter relatively brief descriptions of a variety of clustering methods not obviously members of any of the classes discussed in previous chapters, have been given. Few examples of their use in practice can be found, possibly because of their lack of availability in the major pieces of clustering software— see Appendix.   There are, no doubt, a large number of other clustering techniques that have been developed which have not been described, but to produce a comprehensive review would be a formidable task simply because of the size of the literature and the variety of possible sources.

Complete Linkage on Years

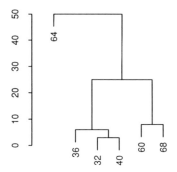

Complete Linkage on States

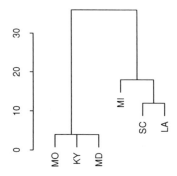

Figure 7.4 Complete linkage dendrograms for the distance matrices in Table 7.5.

Table 7.5 Republican votes for President, and associated distance matrices

(i) Votes

			Year			
State	32	36	40	60	64	68
SC	2	1	4	49	59	39
MI	4	3	4	25	87	14
LA	7	11	14	29	57	23
KY	40	42	40	54	36	44
MD	36	41	37	46	35	42
MO	35	48	38	50	36	45

(ii) Distances between years

	32	36	40	60	64
36	6				
40	3	5			
60	25	24	23		
64	50	47	45	30	
68	18	18	18	8	34

(iii) Distances between states

	SC	MI	LA	KY	MD
MI	18				
LA	12	14			
KY	23	36	26		
MD	28	35	21	4	
MO	28	36	25	4	4

8

Some Final Comments and Guidelines

8.1 Introduction

It should by now be obvious to most readers that the use of cluster analysis in practice does not involve simply the application of one particular technique to the data under investigation, but rather necessitates a series of steps each of which may be dependent on the results of the preceding one. It is generally impossible *a priori* to anticipate what combination of variables, similarity measures and clustering techniques are likely to lead to interesting and informative classifications. Consequently the analysis proceeds through several stages with the researcher intervening if necessary to alter variables, choose a different similarity measure, concentrate on a particular sub-set of individuals etc. The final, extremely important stage, concerns the evaluation of the clustering solution(s) obtained. Are the clusters real or merely artifacts of the algorithms? Do other solutions exist which are better? Can the clusters be given a convincing interpretation? A long list of such questions might be posed.

There is no optimal strategy for either applying clustering or evaluating results, but some suggestions which might be helpful in many situations, are discussed in this chapter.

8.2 Using clustering techniques in practice

Prior to applying any clustering method some graphical representation of the data should be obtained. A number of possibilities were discussed in Chapter 2. Classical principal components analysis is commonly employed to obtain a low-dimensional mapping of the data, although this does not guarantee that the best variables for clustering will be chosen. Other, potentially more useful, ordinations may be obtained from other methods, for example those described by Sammon (1969) and the *projection pursuit* techniques discussed by Jones and Sibson (1987). In addition multidimensional scaling techniques may be used to extract similar visual displays from a calculated proximity matrix. Some authors suggest that if these displays produce no evidence of clustering in the data, then more formal clustering procedures are not required—see, for example, Chatfield and Collins (1980).

The next stage involves deciding whether the raw data or data derived from them should be used as input for the cluster analysis. Various factors govern this choice, such as the clustering technique to be used and the type of raw data concerned, i.e. continuous variables, dichotomous variables, etc. With continuous variables a choice of how to standardize must often be made. The problems involved have been discussed in previous chapters. When using dichotomous variables a choice between the large number of similarity coefficients available often has to be made. The major problem frequently concerns what to do about 'negative' matches. The issues are fully discussed in Sneath and Sokal (1973). Some reduction in number of variables is often called for to make the clustering procedure practically feasible. Scores on a small number of principal components are generally used.

After deciding on the variables to be used, it becomes necessary to choose the clustering technique that is to be applied. By now it should be clear that no one method can be judged to be 'best' in all circumstances. The various studies that have compared a variety of clustering procedures on artificially generated data, all point to the existence of a 'method×data type' interaction. In other words, particular methods will be best for particular types of data. Nevertheless some general recommendations can be made about techniques likely to be useful in the widest range of situations. Amongst the hierarchical class, for example, Ward's method and group average have been found to perform relatively well. Of the optimization methods, minimization of $\det(\mathbf{W})$ has several advantages, and the fitting of mixture densities as described in Chapter 6 does provide the clustering process with a sensible, underlying statistical model.

In many applications it might be reasonable to apply a number of clustering methods. If all produce very similar solutions, the investigator might perhaps have more confidence that the results are worthy of further investigation. Widely different solutions might be taken as evidence against any clear-cut cluster structure. Comparison of different classifications is clearly of some importance here, a topic to be discussed in Section 8.4.

8.3   Assessing clustering solutions

Interpreting the results from a clustering algorithm is often dominated by personal intuition and insight. If the investigator can make sense of the clusters produced, then the cluster analysis is frequently deemed to have to be a success. This may however, be unsatisfactory, particularly in those cases where the user of a clustering algorithm is unsure about the data and has little experience with the particular clustering method being employed. Objective methods of separating artifacts from real structure would clearly be welcome when evaluating a solution. The problem is that clustering algorithms may generate clusters even when applied to random data, and it clearly becomes necessary to try to guard against elaborate interpretation of the solutions found in such cases. In part this is what motivates the development of procedures for testing for number of groups (see Chapters 4, 5 and 6), but here a more general null hypothesis is of interest, that of no structure at all in the data. Rejecting such a hypothesis would not necessarily mean that a clustering structure was appropriate, but it would be a brave (and perhaps ill-advised) researcher who would impose such a structure on data shown to be random.

A number of ways of formulating the 'absence of structure' hypothesis have been proposed, the three most common being

H_0: All $n \times n$ rank order proximity matrices are equally likely

H_0: All permutations of the labels on the n individuals are equally likely

H_0: All sets of n locations in some region of a p-dimensional space are equally likely.

A number of tests for assessing each of these hypotheses have been developed, for example those of Fillenbaum and Rapoport (1971), Ling (1973) and Baker and Hubert (1976). A comprehensive account of such tests is given in Jain and Dubes (1988). Many of the tests have low power against some alternative hypotheses, and according to Gordon (1987), can be markedly influenced by a small number of outliers. Gordon (1980) makes the further very relevant point that these tests are unlikely to find widespread acceptance in a strict hypothesis-testing sense, because of the difficulty of anticipating the behaviour of relevant statistics under the great diversity of different structures which could be present in the data. Nevertheless the procedures are not used as often as they should be, a point partly explained by their absence in the computer packages most used for applying clustering in practice (see Appendix).

The stability of a clustering solution can be examined by randomly dividing the data into two sub-sets and performing an analysis on each subset separately. Similar solutions should be obtained from both sets when the data is clearly structured. Similarly, the analysis may be repeated using only subsets of variables and the results compared. Deletion of a small number of variables from the analysis should not, in most cases, greatly alter the clusters found, if these clusters are 'real' and not mere artifacts of the particular technique being used. (This approach was used quite successfully by Jolliffe *et al.*, 1982). A further procedure which should be applied routinely is to compare the clusters found on variables of interest *other* than those included in the original analysis. If differences between clusters persist with respect to these variables, then this gives some evidence that a useful solution has been found, in the sense that by stating that a particular individual belongs to a certain cluster, information on variables other than those used to derive the cluster is conveyed. (This point has been discussed previously in Chapter 1).

Cohen *et al.* (1977) describe a number of useful graphical techniques for evaluating cluster analysis solutions. The first of these involves consideration of the relative tightness of a k-point group (a potential cluster) compared to other k-point neighbourhoods in the data. The $k - 1$ closest neighbours of

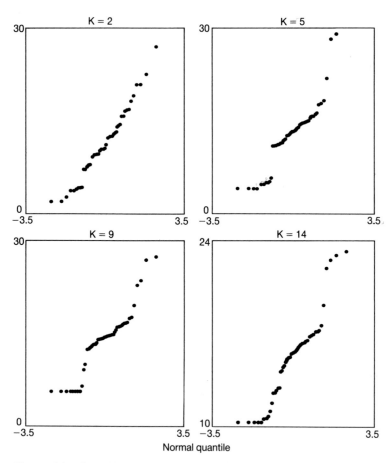

Figure 8.1   Quantile-quantile plots to display clusters. (Taken with permission from Cohen *et al.*, 1977).

each data point are found, and then the average interpoint distance, d_i, among these k individuals over all $k(k-1)/2$ pairs determined. Data points contained in 'real' k-point clusters should give d_i values substantially smaller than data points not in such groupings. Cohen *et al.* suggest comparing the d_is for a given k by means of a normal probability plot. A tight cluster of size j should produce j points with nearly equal d_js which are well separated from the others at the bottom of the plot for $k = j$. Figure 8.1 shows such plots for a data set, using $k = 2, 5, 9$ and 14. The behaviour at the lower ends of these plots suggests the existence of a group of size approximately $k = 9$ in these data.

A further plot described by Cohen *et al.*, is that of squared distances from certain cluster centroids to individuals that are near the centroid. This is useful

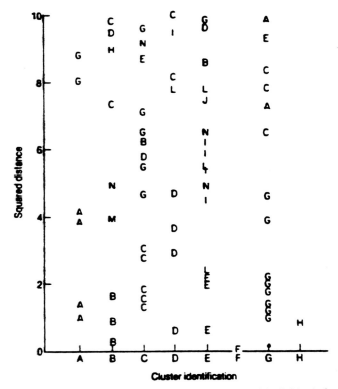

Figure 8.2 Plot of squared distances of selected individuals from their cluster centroids. (Taken with permission from Cohen *et al.*, 1977).

for examining the internal cohesiveness of a cluster. For each cluster centroid, the distances of every individual from that centroid are plotted above the cluster identification shown on the x-axis. The symbol plotted corresponds to the cluster in which the individual resides. An example of such a plot is seen in Figure 8.2. From this point it is clear that clusters F and H are very well separated from neighbouring individuals, that an individual not assigned to cluster E is still reasonably close to it and that there is a large distance between the members of cluster A furthest from their centroid and the next closest individuals not assigned to A.

Another simple graphical aid for evaluating clustering solutions, again suggested by Cohen *et al.*, can be used for examining the clusters in terms of either variables used to form the clusters or other variables of interest. Here the clusters are again identified along the x-axis, and above each label the values on a particular variable are plotted, for each individual in the cluster. The median of the cluster is also plotted. The plot can then be used to compare individuals in the same cluster on the variable in question, and to

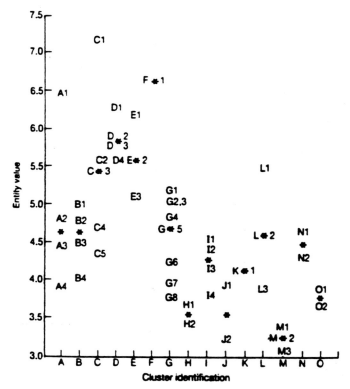

Figure 8.3 Plot of values of a single variable for selected individuals in various clusters. (Taken with permission from Cohen *et al.*, 1977).

make multiple comparisons across clusters. An example of such a plot appears in Figure 8.3. This shows that clusters A and B are quite similar on this variable apart from one individual, A1 in cluster A. Cluster E tends to contain individuals with large values on this variable and these have moderate spread, whilst cluster M has much smaller values with small spread.

8.4   Comparing classifications

Often when carrying out a cluster analysis of a set of multivariate data, it may be necessary to compare two or more clusterings of the same set of individuals. The solutions to be compared might arise from the use of different clustering methods on the same data set, or from the same clustering method applied to different similarity or distance matrices arising from the raw data. A further possibility is that the different solutions arise from applying the same clustering procedure to the same proximity matrix derived from data sets from different sources. For small or moderate sized examples, it might be possible to make the required comparisons informally, simply by inspecting the clusterings to determine the important clusters and/or by examining dendrograms and assessing where they match and where they differ. In many practical applications however, such an approach is likely to be extremely laborious and time-consuming and it also affords no measurement as to the similarity of the various solutions. Consequently a number of authors have suggested more formal procedures for comparing classifications. One of the most commonly used of these is an index, R_g, suggested by Rand (1971), which may be defined as follows. Let n be the total number of individuals to be clustered. Then for a given number of clusters, g, R_g is defined as the ratio of the sum of the number of pairs of objects that cluster together in the two clusterings under comparison and the number of pairs of objects that fall in different clusters in both clusterings, to the total number of pairs, $\binom{n}{2}$. Thus R_g can be interpreted as the probability that two individuals are treated alike in both solutions. Fowlkes and Mallows (1983) show that R_g may be written as

$$R_g = \left[T_g - \frac{1}{2}P_g - \frac{1}{2}Q_g + \binom{n}{2} \right] \Big/ \binom{n}{2} \tag{8.1}$$

where

$$T_g = \sum_{i=1}^{g} \sum_{i=1}^{g} m_{ij}^2 - n \tag{8.2}$$

$$P_g = \sum_{i=1}^{g} m_{i\cdot}^2 - n \tag{8.3}$$

$$Q_g = \sum_{i=1}^{g} m_{\cdot j}^2 - n \tag{8.4}$$

and the quantity m_{ij} is the number of individuals in common between the ith cluster of the first solution, and the jth cluster of the second (the clusters in the two solutions may each be labelled arbitrarily from 1 to g). The terms $m_{i\cdot}$ and $m_{\cdot j}$ are appropriate marginal totals of the matrix of m_{ij} values.

R_g lies in the interval $(0,1)$, and takes its upper limit when there is complete agreement between the two classifications. The properties of this index have been discussed by Rand (1971) and Fowlkes and Mallows (1983) but here it is sufficient to give an example of its use in practice. For this, two classifications of the iris data (see Chapters 2 and 6) will be compared. Both arise from applying a three component multivariate normal mixture, the first assuming that each component has the same covariance matrix, the second allowing different covariance matrices. The required matrix of m_{ij} values is

Unequal Cov. Matrices	Equal Covariance Matrices		
	1	2	3
1	0	0	50
M = 2	2	47	0
3	36	15	0

The various terms needed to calculate the Rand index are

$$T_g = 6234 - 150$$
$$P_g = 5252 - 150$$
$$Q_g = 7788 - 150$$

leading to

$$R_g = [6234 - 150 - \frac{1}{2}(5252 - 150) - \frac{1}{2}(7788 - 150) + 11175]/11175$$
$$= 0.97$$

Here the agreement between the two classifications is clearly very good.

Fowlkes and Mallows (1983) describe a method for comparing two hierarchical clusterings which uses an index which is similar in some respects to that of Rand. The measure, B_g is defined as

$$B_g = T_g/\sqrt{P_g Q_g} \tag{8.5}$$

but instead of simply using the value of the index for a particular number of groups, Fowlkes and Mallows suggest plotting $(g, B_g), g = 2, \ldots, n - 1$, for each pair of partitions obtained from the two dendrograms. A series of Monte Carlo studies reveal the potential of this procedure for comparing classifications. Additionally the plots appear to have potential for selecting the appropriate number of clusters.

Other indices for comparing the results from different classifications of a data set are described by Johnson (1968), Anderberg (1973) and Hubert and Levine (1976). Arabie and Boorman (1973) used multidimensional scaling to study the similarity between the measures themselves.

8.5   Summary

The methods of cluster analysis can be valuable tools in the exploration of multivariate data. By organizing such data into subgroups or clusters, clustering may help the investigator discover the characteristics of any structure or pattern present. Applying the methods in practice however, requires considerable care if over-interpretation of the solutions obtained is to be avoided. Much attention needs to be given to questions of cluster validity, although such questions are rarely straightforward and are full of traps for the unwary (see Dubes and Jain, 1979). Simply applying a particular method of cluster analysis to a data set and accepting the solution at face value is, in general, not adequate.

Searching for a comment to end this book, I first returned to Cormack's highly influential review of clustering techniques published over twenty years ago. Cormack was almost wholly negative about the practical usefulness of these techniques—the following quotation from the paper gives the flavour of Cormack's point of view

"The summarization of large quantities of multivariate data by clusters, undefined *a priori*, is increasingly practised, often irrelevantly and unjustifiably. The availability of computer packages of classification techniques has led to the waste of more valuable time than any other statistical innovation."

Should such a quotation be used as a final *caveat emptor*? On reflection I thought not. Although Cormack undoubtedly had some justification for expressing such reservations about clustering, many of his criticisms went too far. Consequently I sought a little more balanced and perhaps slightly more positive comment on which to end. This I found in a recent paper by Hansen and Tukey (1992). The following quotation summarizes nicely how sensible researchers should view the clustering process:

"What purpose do clusters serve? Presumably purposes of description, where they may help with either graphical or verbal description, and, to a degree, purposes of prediction or prognosis—as where we try to separate persons with an upper respiratory infection (head cold) from those who do not, even though there is a continuous gradation, in some patients, from one state to the other. If we can avoid asking too much of clustering techniques, they can serve us better."

Appendix A
Software

A.1 Introduction

The application of most methods of cluster analysis depends on the availability
of suitable computer software. Methods are only likely to be used routinely
if they exist in well documented and easily used packages. Consequently it is
such packages which are featured in this Appendix.

Many researchers with a cluster analysis application are likely to begin their
search for suitable software with one of the three main statistical packages,
BMDP, SPSS and SAS. Each of these provides the means for the practical use
of many of the methods described in this text. None is however, ideal for this
purpose.

A.2 BMDP

The BMDP package contains a number of programs for cluster analysis but
this area is perhaps the weakest part of what is otherwise excellent statistical
software. The programs provided are as follows:

A.2.1 2M

This program implements single linkage or centroid hierarchical clustering
using one of eleven distance measures.

A.2.2 KM

This program essentially implements a 'k-means' type algorithm as described
in Chapter 5. Solutions provide a partition of the individuals into clusters such
that each individual belongs to the cluster whose centre is closest in Euclidean
distance terms.

A.2.3 3M

This program constructs clusters of individuals that are alike for subsets of
variables. It implements the method of Hartigan (1975) described in Chapter 7.

A.2.4 1M

Here the emphasis is on the clustering of variables by way of a hierarchical
scheme.

In addition to these specialized programs for clustering, BMDP contains programs for principal components and correspondence analysis which are often usefully employed prior to the cluster analysis proper.

A.3   SPSS

This package provides two programs relevant to cluster analysis, PROXIMITIES and CLUSTER. The first computes a variety of similarity and distance measures and the second allows the agglomerative hierarchical techniques described in Chapter 4 to be applied. Again cluster analysis is not a strong feature of this software.

In addition SPSS can perform principal components, correspondence analysis and multidimensional scaling (see Chapter 3).

A.4   SAS

The SAS package contains a variety of clustering procedures. They are not particularly well integrated, however, nor well described in the manual. The methods available are:

A.4.1   CLUSTER

Performs hierarchical clustering using the centroid method, Ward's method, or average linkage on squared Euclidean distances.

A.4.2   FASTCLUS

Finds clusters using a 'k-means' algorithm.

A.4.3   OVERCLUS

Finds overlapping clusters from similarity data.

Additionally the package can be used to perform principal components analysis and multidimensional scaling.

A.5   CLUSTAN

CLUSTAN is a computer package designed solely for the purposes of clustering. It was first developed in the 1960s and although it has been through several versions since then, the latest being 1987, in many respects it looks a little dated. Nevertheless it implements many of the clustering techniques described in this book, including several not available in any other package. A wide variety of distance and similarity measures may be used and the clustering methods available include agglomerative hierarchical methods, divisive methods, k-means, minimization of trace($\mathbf{W}$) and det($\mathbf{W}$), Sibson and Jardine's overlapping cluster procedure and multivariate normal mixtures. A selection of associated techniques such as principal components analysis are also included in the package.

Information about the package is available from

CLUSTAN Ltd., 16 Kingsburgh Rd., Edinburgh EH12 6DZ, Scotland.

A.6 SYSTAT

In SYSTAT clustering is available through the commands JOIN and KMEANS. The first implements hierarchical clustering, the second a 'k-means' algorithm. In addition SYSTAT has a variety of powerful graphical facilities which allow the initial exploration of multivariate data along the lines described in Chapter 2. Details of SYSTAT are obtainable from

SYSTAT Inc., 1800 Sherman Avenue, Evanston, Illinois 60201-3793, USA or 47 Hartfield Crescent, West Wickham, Kent BR4 9DW.

A.7 NTSYS

NTSYS is a collection of programs for finding and displaying patterns and structure in multivariate data. Several hierarchical clustering methods are implemented and a variety of proximity measures are available. The package also provides non-metric multidimensional scaling and correspondence analysis.

Details are available from

F. James Rohlf, Dept of Ecology and Evolution, State University of New York, Stony Brook, N.Y. 11794, USA.

The PC version of the program is distributed by

Exeter Software, 100 North Country Road, Bldg. B, Setauket, NY 11733.

A.8 CLUSTAR

CLUSTAR performs hierarchical cluster analysis and provides a variety of methods of standardization and similarity measures. Details are obtainable from

H. Charles Romesburg, College of Natural Resources, Utah State University, Logan, UT 84322-5200, USA.

In addition to specialised clustering packages, more general software such as GENSTAT 5 and S-PLUS provide some clustering facilities. The former, for example, has extensive facilities for calculating similarity and distance measures, and for hierarchical clustering and non-hierarchical clustering using one or other of the criteria mentioned in Chapter 5. S-PLUS has procedures for calculating a number of distance and similarity measures and for applying particular hierarchical cluster analysis methods. The main strengths of both GENSTAT and S-PLUS however, are (i) that they allow the investigator exploring complex multivariate data access to extensive graphical techniques and (ii) they provide very powerful programming languages for implementing non-standard analyses.

References

Adamson, G.W. and Bawden, D. (1981). Comparison of hierarchical cluster analysis techniques for automatic classification of chemical structures. *J. Chem. Inf. Comput. Sci*, **21**, 204–209.

Aitkin, M., Anderson, D., and Hinde, J. (1981). Statistical modelling of data on teaching styles. *J. Roy. Stat. Soc.*, A., **144**, 419–448.

Anderberg, M.R. (1973). *Cluster Analysis for Applications*, Academic Press, New York.

Anderson, A.J.B. (1971). Ordination methods in ecology. *J. Ecol.*, **59**, 713–726.

Anderson, J.J. (1985). Normal mixtures and the number of clusters problem. *Comput. Statist. Quarterly*, **2**, 3–14.

Andrews, D.F. (1972). Plots of high dimensional data. *Biometrics*, **28**, 125–136.

Arabie, P. and Boorman, S.A. (1973). Multidimensional Scaling of measures of distance between partitions. *J. Math. Psychol.*, **10**, 148–203.

Baker, F.B. (1974). Stability of two hierarchical grouping techniques—Case 1. Sensitivity to data errors. *J. Amer. Statist. Assoc.*, **69**, 440–445.

Baker, F.B. and Hubert, L.J. (1976). A graph theoretic approach to goodness-of-fit in complete link hierarchical clustering. *J. Amer. Statist. Assoc.*, **71**, 870–878.

Ball, G.H. and Hall, D.J. (1967). A clustering technique for summarizing multivariate data. *Behavl. Sci.*, **12**, 153–155.

Ballard, J. and Gothfredson, D. (1963). *Predictive Attribute Analysis and Prediction of Performance*. Social Agency Effectiveness Study, California Medical Faculty, Vacaville, California.

Beale, E.M.L. (1969). Euclidean cluster analysis. *Bull. I.S.I.*, **43**, Book 2, 92–94.

Binder, D.A. (1978). Bayesian cluster analysis, *Biometrika*, **65**, 31–38.

Blashfield, R.K. (1976). Mixture model tests of cluster analysis. Accuracy of four agglomerative hierarchical methods. *Psychol. Bulletin*, **83**, 377–385.

Bock, H.H. (1980). Simultaneous clustering of objects and variables. In *Recherche d'Informatique et d'Automatique* (R. Tomasone, ed), Le Chesney, France.

Bonner, R.E. (1964). On some clustering techniques. *I.B.M.J. Res Dev.*, **8**, 22–32.

Boyce, A.J. (1969). Mapping diversity. A comparative study of some numerical methods. In *Numerical Taxonomy* (A.J. Cole, ed), Academic Press, New York.

Calinski, T. and Harabasz, J. (1974). A dendrite method for cluster analysis. *Communications in Statistics*, **3**, 1–27.

Carmichael, J.W. and Sneath, P.H.A. (1969). Taxometric Maps. *Syst. Zool.*, **18**, 402–415.

Carmichael, J.W., George, J.A. and Julius, R.S. (1968). Finding natural clusters. *Syst. Zool.*, **17**, 144–150.

Cattell, R.B. and Coulter, M.A. (1966). Principles of behavioural taxonomy and the mathematical basis of the taxonome computer program. *Br. J. Math. Statist. Psychol.*, **19**, 237–269.

Chambers, J.M., Cleveland, W.S., Kleiner, B. and Tukey, P.A. (1983). *Graphical Methods for Data Analysis*, Wadsworth, Belmont, California.

Chandon, J.L., Lemaire, J. and Pouget, J. (1980). Construction de l'utrametrique la plus proche d'une dissimilarité au sens des mondres carres. R.A.I.R.O. *Recherche Operationnelles*, **14**, 157–170.

Chang, W.C. (1983). On using principal components before separating a mixture of multivariate normal distributions. *Applied Statistics*, **32**, 267–275.

Chatfield, C. and Collins, A.J. (1980). *Introduction to Multivariate Analysis*, Chapman & Hall, London.

Cheetham, A.H. and Hazel, M. (1969). Binary (presence-absence) similarity coefficients. *J. of Paleontology*, **43**, 1130–1136.

Chernoff, H. (1973). The use of faces to represent points in *k*-dimensional space graphically. *J. Amer. Stat. Assoc.*, **68**, 361–368.

Clark, V.A., Chapman, J.M., Coulson, A.H. and Hasselblad, V. (1968). Dividing the blood pressures from the Los Angeles heart study into two normal populations. *Johns Hopkins Medical Journal*, **122**, 77–83.

Clifford, H.T. and Stephenson, W. (1975). *An Introduction to Numerical Classification*, Academic Press, New York.

Cohen, A., Gnanadesikan, R., Kettenring, J.R. and Landwehr, J.M. (1977). Methodological developments in some applications of clustering. In *Applications of Statistics* (P.R. Krishnaiah, ed), North Holland Publishing Co., Amsterdam.

Cohen, J. (1960). A coefficient of agreement for normal scales. *Educ. and Psychol. Measur.*, **20**, 37–46.

Constantine, A.G. and Gower, J.C. (1978). Graphical representation of asymmetric matrices. *Applied Statistics*, **27**, 297–304.

Cole, A.J. and Wishart, D. (1970). An improved algorithm for the Jardine–Sibson method of generating overlapping clusters. *Comp. J.*, **13**, 156-163.

Cooper, P.W. (1964). Non supervised adaptive signal detection and pattern recognition. *Information and Control*, **7**, 416–444.

Cormack, R.M. (1971). A review of classification. *J. Roy. Statist. Soc.*, A, **134**, 321–367.

Coste, J., Spira, A., Ducimetiere, P. and Paolaggi, J.B. (1991). Clinical and psychological diversity of non-specific low-back pain. A new approach towards the classification of clinical subgroups. *J. Clin. Epidemiol*, **44**, 1233–1245.

Cunningham, K.M. and Ogilvie, J.C. (1972). Evaluation of hierarchical grouping techniques: a preliminary study. *Comp. J.*, **15**, 209–213.

Czekanowski, J. (1909). Zir Differential-diagnose de Neadertalgruppe. Korrespondenzblatt der Deutschen Geselschaft fur Anthropologie. *Ethnologie and Urgeschichte*, **40**, 44–47.

Czekanowski, J. (1932). Coefficient of racial likeness and durchschnittliche differenz. *Anthropologischer Anzeiger*, **9**, 227–49.

Davenport, M. and Studdert-Kennedy, G. (1972). The statistical analysis of aesthetic judgement: an exploration. *Applied Statistics*, **21**, 324–332.

Dawkins, B. (1989). Multivariate analysis of national track records. *American Statistician*, **43**, 110–115.

Day, N.E. (1969). Estimating the components of a mixture of normal distributions. *Biometrika*, **56**, 463–474.

Defrays, D. (1977). An efficient algorithm for the complete link method. *Computer J.*, **20**, 364–366.

Dempster, A.P., Laird, N.M. and Rubin D.B. (1977). Maximum likelihood from incomplete data via the EM algorithm. *J. Roy. Statist. Soc.*, B, **39**, 1–38.

De Sarbo, W.S., Carroll, J.D., Clark, L.A. and Green, P.E. (1984). Synthesized clustering: a method for amalgamating alternative clustering bases with differential weighting of variables. *Psychometrika*, **49**, 57–78.

De Soete, G. (1984). A least squares algorithm for fitting an ultrametric tree to a dissimilarity matrix. *Pattern Recognition Letters*, **2**, 133–137.

Dick, N.P. and Bowden, D.C. (1973). Maximum likelihood estimation for mixtures of two normal distributions. *Biometrics*, **29**, 781–790.

Dixon, J.K. (1979). Pattern recognition with missing data. *IEEE Transactions on Systems, Man and Cybernetics*, SMC9, 617–621.

Dubes, R. and Jain, A.K. (1979). Validity studies in clustering methodologies. *Pattern Recognition*, **8**, 247–260.

Duda, R.O. and Hart, P.E. (1973). *Pattern Classification and Scene Analysis*. John Wiley & Sons, New York.

Duflou, H., Maenhaut, W. and De Reuck, J. (1990). Application of principal component and cluster analysis to the study of the distribution of minor and trace elements in normal human brain. *Chemometrics and Intelligent Laboratory Systems*, **9**, 273–286.

Dunn, G. and Everitt, B.S. (1982). *An Introduction to Mathematical Taxonomy*, Cambridge University Press, Cambridge.

Embrechts, P. and Herzberg, A.M. (1991). Variations of Andrews' plots. *International Statistical Review*, **59**, 175–194.

Everitt, B.S. (1981). A Monte Carlo investigation of the likelihood ratio test for the number of components in a mixture of normal distributions. *Multiv. Behav. Res.*, **16**, 171–180.

Everitt, B.S. (1987). *Introduction to optimization methods and their application in Statistics*, Chapman & Hall, London.

Everitt, B.S. (1988). A finite mixture model for the clustering of mixed mode data. *Statistics and Probability Letters*, **6**, 305–309.

Everitt, B.S. and Dunn, G. (1988). Log-linear modelling, latent class analysis, or correspondence analysis. Which method should be used for the analysis of categorical data? In *Latent Trait and Latent Class Models* (R. Langeheine and J. Rost, ed), Plenum Press, New York.

Everitt, B.S. and Dunn, G. (1991). *Applied Multivariate Data Analysis*, Edward Arnold, London.

Everitt, B.S. and Hand, D.J. (1981). *Finite Mixture Distributions*, Chapman & Hall, London.

Everitt, B.S. and Merette, C. (1989). The clustering of mixed-mode data: a comparison of possible approaches. *Journal of Applied Statistics*, **17**, 283–297.

Everitt, B.S., Gourlay, A.J. and Kendell, R.E. (1971). An attempt at validation of traditional psychiatric syndromes by cluster analysis. *Br. J. Psychiat.*, **119**, 399–412.

Farmer, A. and McGuffin, P. (1989). The Classification of the depressions. Contemporary confusion revisited. *Brit. J. Psych*, **155**, 437–443.

Farmer, A., McGuffin, P. and Spitznagel, E. L. (1983). Heterogeneity in schizophrenia: a cluster analytic approach. *Psychiatric research*, **8**, 1–12.

Ferligoj, A. and Batagelj, V. (1982). Some types of clustering with relational constraints. *Psychometrika*, **47**, 541–552.

Fillenbaum, S., and Rapoport, A (1971). *Structures in the Subjective Lexicon*, Academic Press, New York.

Fisher, L. and Van Ness, J.W. (1973). Admissible clustering procedures. *Biometrika*, **58**, 91–104.

Fisher, W.D. (1958). On grouping for maximum homogeneity. *J. Amer. Statist. Assoc.*, **53**, 789–798.

Fleiss, J.L. and Zubin, J. (1969). On the methods and theory of clustering. *Multivariate Behaviour Res.*, **4**, 235–250.

Florek, K., Lukaszewiez, J., Perkal, J., Steinhaus, H. and Zubrzchi, S. (1951). Sur la liason et la division des points d'un ensemble fini. *Colloquium Mathematicum*, **2**, 282–285.

Flury, B. and Riedwyl, H. (1981). Graphical representation of multivariate data by means of asymmetrical faces. *J. Amer. Statist. Assoc.*, **76**, 757–765.

Forgey, E.W. (1965). Cluster analysis of multivariate data: efficiency versus interpretability of classification. *Biometrics*, **21**, 768–769.

Fowlkes, E.B. and Mallows, C.L. (1983). A method for comparing two hierarchical clusterings. *J. Amer. Statist. Assoc.*, **78**, 553–569.

Friedman, H.P. and Rubin, J. (1967). On some invariant criteria for grouping data. *J. Amer. Statist. Assoc.*, **62**, 1159–1178.

Gabriel, K.R. (1981). Biplot display of multivariate matrices for inspection of data and diagnosis. In *Interpreting Multivariate Data* (V. Barnett, ed), John Wiley & Sons, London.

Gitman, I. and Levine, M.D. (1970). An algorithm for detecting unimodal fuzzy sets and its application as a clustering technique. *IEE Trans. Comp.*, C19, 583–593.

Gnanadesikan, R. (1977). *Methods for Statistical Data Analysis of Multivariate Observations*, John Wiley & Sons, New York.

Goodman, L.A. (1974). Exploratory latent structure analysis using both identifiable and unidentifiable models. *Biometrika*, **61**, 215–231.

Gordon, A.D. (1973). Classification in the presence of constraints. *Biometrics*, **29**, 821–827.

Gordon, A.D. (1980). *Classification*, Chapman & Hall, London.

Gordon, A.D. (1987). A review of hierarchical classification. *J. Roy. Statist. Soc.* **150**, 119–137.

Gordon, A.D. (1992). Hierarchical Classification, in *Clustering and Classification* (P. Arabie, L. Hubert, and G. De Soete, eds), World Scientific Press.

Gordon, A.D. and Henderson, J.T. (1977). An algorithm for Euclidean sum of squares classification. *Biometrics*, **33**, 355–362.

Gowda, K.C. and Krishna, G. (1978). Agglomerative clustering using the concept of mutual nearest neighbourhood. *Pattern Recognition*, **10**, 105–112.

Gower, J.C. (1966). Some distance properties of latent root and vector methods used in multivariate analysis. *Biometrika*, **53**, 325–338.

Gower, J.C. (1967a). Multivariate analysis and multidimensional geometry. *The Statistician*, **17**, 13–25.

Gower, J.C. (1967b). A comparison of some methods of cluster analysis. *Biometrics*, **23**, 623–628.

Gower, J.C. (1971). A general coefficient of similarity and some of its properties. *Biometrics*, **27**, 857–872.

Gower, J.C. (1975). Goodness-of-fit criteria for classification and other patterned structures. In *Proc. of the 8th International Conf. on Numerical Taxonomy*, 38–62.

Gower, J.C. (1985). Measures of similarity, dissimilarity and distance. In *Encyclopedia of Statistical Sciences, Volume 5* (S. Kotz, N.L. Johnson and C.B. Read, eds), John Wiley & Sons, New York.

Gower, J.C. (1988). Classification, geometry and data analysis. In *Classification and Related Methods of Data Analysis* (H.H. Bock, ed), Elsevier, North-Holland, Amsterdam.

Gower, J.C. and Ross, G.J.S. (1969). Minimum spanning trees and single linkage cluster analysis. *Appl. Statist.*, **18**, 54–64.

Green, P.E., Frank, R.E. and Robinson, P.J. (1967). Cluster analysis in test market selection. *Management Science*, **13**, 387–400.

Greenacre, M. (1984). *Theory and Applications of Correspondence Analysis*, Academic Press, Florida.

Haltenorth, T. (1937). Die verwandschaftliche Stellung der Grosskatzen zu eviander. *Zeitschrift fur Saugetierkunde*, **12**, 97–240.

Hand, D.J. (1981). *Discrimination and Classification*, John Wiley & Sons, Chichester.

Hands, S. and Everitt, B.S. (1987). A Monte Carlo study of the recovery of cluster structure in binary data by hierarchical clustering techniques. *Multiv. Behav. Res.*, **22**, 235–243.

Hansen, K.M. and Tukey, J.W. (1992). Tuning a major part of a clustering algorithm. *International Statistical Review*, **60**, 21–44.

Hartigan, J.A. (1967). Representation of similarity matrices by trees. *J. Amer. Statist. Assoc.*, **62**, 1140–1158.

Hartigan, J.A. (1972). Direct clustering of a data matrix. *J. Amer. Statist. Assoc.*, **67**, 123–129.

Hartigan, J.A. (1975). *Clustering Algorithms*, John Wiley & Sons, New York.

Hasselblad, V. (1966). Estimation of parameters for a mixture of normal distributions. *Technometrics*, **8**, 431–444.

Hawkins, D.M., Muller, M.W. and ten Krooden, J.A. (1982). Cluster Analysis. In *Topics in Applied Multivariate Analysis* (D.M. Hawkins, ed), Cambridge University Press, Cambridge.

Heinrich, I., O'Hara, H., Sweetman, B. and Anderson, J.A.D. (1985). Validation aspects of an empirically derived classification for non-specific low back pain. *The Statistician*, **34**, 215–230.

Hernandez-Avilia, A. (1979). *Problems in Cluster Analyis*. Unpublished D. Phil. thesis, University of Oxford.

Hodson, F.R. (1971). Numerical typology and prehistoric archaeology. In *Mathematics in the Archaelogical and Historical Sciences* (Hodson, F.R., Kendall, D.G. and Tautu, P.A., eds), Edinburgh: University Press.

Hosmer, D.W. (1973). On the MLE of the parameters of a mixture of two normal distributions when the sample size is small. *Commun. Statist.* **1**, 217–227.

Hubert, L.J. (1974). Approximate evaluation techniques for the single link and complete link hierarchical clustering procedures. *J. Amer. Statist. Assoc.* **69**, 698–704.

Hubert, L.J. and Levine, J.R. (1976). Evaluating object set partitions: free sort analysis and some generalizations. *J. Verbal Learning and Verbal Behaviour*, **15**, 549–570.

Ihm, P. (1965). Automatic classification in anthropology. In *The Use of Computers in Anthropology* (D. Hymes, ed), Mouton & Co., The Hague.

Ishmail, M.A. and Kamel, M.S. (1989). Multidimensional data clustering utilizing hybrid search strategies. *Pattern Recognition*, **22**, 75–89.

Jackson, J.E. (1991). *A User's Guide to Principal Components*, John Wiley & Sons, New York.

Jain, A.K. and Dubes, R.C. (1988). *Algorithms for Clustering Data*, Prentice Hall, Englewood Cliff, NJ, USA.

Jancey, R.C. (1966). Multidimensional group analysis. *Aust. J. Bot.*, **14**, 127–130.

Jardine, N. and Sibson, R. (1968). The construction of hierarchic and non-hierarchic classifications. *Comp. J.*, **11**, 117–184.

Jardine, N. and Sibson, R. (1971). *Mathematical Taxonomy*, John Wiley & Sons, Chichester.

Jensen, R.E. (1969). A dynamic programming algorithm for cluster analysis. *Op. Res.*, 1034–1056.

Johnson, L.A.S. (1968). Rainbows end: the quest for an optimal taxonomy. *Proc. Linn. Soc. N.S.W.*, **93**, 8–45.

Johnson, S.C. (1967). Hierarchical clustering schemes. *Psychometrika*, **32**, 241–254.

Jones, K.S. and Jackson, D.M. (1967). Current approaches to classification and clump finding at the Cambridge Language Research Unit. *Comp. J.*, **1**, 29–37.

Jones, M.C. and Sibson, R. (1987). What is projection pursuit? *J. Roy. Statist. Soc. A*, **150**, 1-38.

Jolliffe, I.T. (1986). *Principal Components Analysis*, Springer-Verlag, New York.

Jolliffe, I.T., Jones, B., and Morgan, B.J.T. (1982). Utilising clusters: a case study involving the elderly. *J. Roy. Statist. Soc.* A, **145**, 224–236.

Katz, J.O. and Rohlf, F.J. (1973). Function point cluster analysis. *Systematic Zoology*, **22**, 295–301.

Kaufman, L. and Rousseeuw, P.J. (1990). *Finding Groups in Data*, John Wiley & Sons, New York.

Klein, R.W. and Dubes, R.C. (1989). Experiments in projection and clustering by simulated annealing. *Pattern Recognition*, **22**, 213–220.

Kleiner, B. and Hartigan, J.A. (1981). Representing points in many dimensions by trees and castles. *J. Amer. Statist. Assoc.*, **76**, 260–269.

Koontz, W.L.G., Narendra, P.M. and Fukunaga, K. (1975). A branch and bound clustering algorithm. *IEEE Transactions on Computers*, C-24, 908–915.

Kruskal, J.B. (1964). Multidimensional scaling by optimising goodness of fit to non-metric hypothesis. *Psychometrika*, **29**, 1–27.

Kruskal, J.B. and Wish, M. (1978). *Multidimensional Scaling*, Sage, Newbury Park, CA, USA.

Krzanowski, W.J. (1988). *Principles of Multivariate Analysis: A User's Perspective*, Oxford University Press, Oxford.

Kuiper, F.K. and Fisher, L. (1975). A Monte Carlo comparison of six clustering procedures. *Biometrics*, **31**, 777–783.

Kurtz, A., Moller, H.J., Bavidl, G., Buirk, F., Torhost, A., Wachther, C. and Lauter, H. (1987). Classification of parasuicide by cluster analysis. *Brit. J. Psych.*, **150**, 520–525.

Lambert, J.M. and Williams, W.T. (1962). Multivariate methods in plant ecology, IV. Nodal Analysis. *J. Ecol.*, **50**, 775–802.

Lambert, J.M. and Williams, W.T. (1966). Multivariate methods in plant ecology, V. Comparison of information analysis and association analysis. *J. Ecol.*, **54**, 635–664.

Lance, G.N. and Williams, W.T. (1966). Computer programs for hierarchical polythetic classification. *Comp. J.*, **9**, 60–64.

Lance, G.N. and Williams, W.T. (1967). A general theory of classificatory sorting strategies: 1. Hierarchical systems. *Comp. J.*, **9**, 373–380.

Lazarsfeld, P.L. and Henry, N.W. (1968). *Latent Structure Analysis*. Houghton Mifflin Co., Boston.

Lechevallier, Y. (1980). Classification sous contraintes. In *Optimization en Classification Automatique* (E. Diday ed), INRIA, Paris.

Ling, R.F. (1972). On the theory and construction of *k*-clusters. *Comp. J.*, **15**, 326–332.

Ling, R.F. (1973). Probability theory of cluster analysis. *J. Amer. Statist. Assoc.*, **68**, 159–164.

Little, R.A. and Rubin, D.B. (1987). *Statistical Analysis with Missing Data*. John Wiley & Sons, New York.

Liu, G.L. (1968). *Introduction to Combinatorial Mathematics*, McGraw Hill.

Lorr, M., Klett, C.J. and McNair, D.M. (1963). *Syndromes of Psychosis*. Pergamon, London.

Lukasova, A. (1979). Hierarchical agglomerative clustering procedure. *Pattern Recognition*, **11**, 365–381.

MacNaughton-Smith, P. (1965). *Some Statistical and Other Numerical Techniques for Classifying Individuals*. Home Office Research Unit Report No.6. H.M.S.O., London.

MacNaughton-Smith, P., Williams, W.T., Dale, M.B. and Mockett, L.G. (1964). Dissimilarity analysis. *Nature*, **202**, 1034–1035.

MacQueen, J. (1967). Some methods for classification and analysis of multivariate observations. *Proc. 5th Berkeley Symp.*, **1**, 281–297.

Maronna, R. and Jacovkis, P.M. (1974). Multivariate clustering procedures with variable metrics. *Biometrics*, **30**, 499–505.

Marriott, F.H.C. (1971). Practical problems in a method of cluster analysis. *Biometrics*, **27**, 501–514.

Marriott, F.H.C. (1982). Optimization methods of cluster analysis. *Biometrika*, **69**, 417–421.

McLachlan, G.J. (1992). Cluster analysis and related techniques in medical research. *Statistical Methods in Medical Research*, **1**, 27–48.

McLachlan, G.J. and Basford, K.E. (1988). *Mixture Models: Inference and Applications to Clustering*, Marcel Dekker, New York.

McRae, D.J. (1971). Micka, a Fortran IV iterative *K*-means cluster analysis program. *Behavl. Sci.*, **16**, 423–424.

Michener, C.D. (1970). Diverse approaches to systematics. *Evolutionary Biology*, **4**, 1–38.

Milligan, G.W. (1980). An examination of the effect of six types of error perturbation on fifteen clustering algorithms. *Psychometrika*, **45**, 325–342.

Milligan, G.W. (1989). A study of the beta-flexible clustering method. *Multiv. Behav. Res.*, **24**, 163–176.

Milligan, G.W. and Cooper, M.C. (1985). An examination of procedures for determining the number of clusters in a data set. *Psychometrika*, **50**, 159–179.

Mojena, R. (1977). Hierarchical grouping methods and stopping rules: an evaluation. *Computer J.*, **20**, 359–363.

Morgan, B.J.T. (1973). Cluster analysis of two acoustic confusion matrices. *Perception and Psychophysics*, **13**, 13–24.

Morrison, D.G. (1967). Measurement problems in cluster analysis. *Management Science*, **13**, 775–780.

Murtagh, F. (1985). *Multidimensional clustering algorithms*. COMPSTAT Lectures 4. Physica-Verlag, Vienna.

Needham, R.M. (1965). Computer methods for classification and grouping. In *The Use of Computers in Anthropology* (D. Hymes, ed), 345–356, Mouton & Co., The Hague.

Needham, R.M. (1967). Automatic classification in linguistics. *The Statistician*, **17**, 45–54.

Parker-Rhodes, A.F. and Jackson, D.M. (1969). Automatic classification in the ecology of the higher fungi. In *Numerical Taxonomy* (A.J. Cole, ed), Academic Press, New York.

Paykel, E.S. (1971). Classification of depressed patients: a cluster analysis derived grouping. *Brit. J. Psychiatry*, **118**, 275–288.

Paykel, E.S. and Rassaby, E. (1978). Classification of suicide attempters by cluster analysis. *Brit. J. of Psychiatry*, **133**, 42–52.

Pearson, K. (1894). Contribution to the mathematical theory of evolution. *Phil. Trans.* A, **185**, 71–110.

Pickering, G.W. (1961). *The Nature of Essential Hypertension*. Grune and Stratton, New York.

Pickering, R.M. and Forbes, J.F. (1984). A classification of Scottish infants using latent class analysis. *Statistics in Medicine*, **3**, 249–259.

Pielou, E.C. (1969). *An Introduction to Mathematical Ecology*, John Wiley & Sons, New York.

Pilowsky, I., Levine, S. and Boulton, D.M. (1969). The classification of depression by numerical taxonomy. *Br. J. Psychiat.*, **115**, 937–945.

Platt, R. (1959). The nature of essential hypertension. *Lancet*, **55**, 1092.

Platt, R. (1963). Heridity in hypertension, *Lancet*, **59**, 899.

Powell, G.E., Clark, E. and Bailey, S. (1979). Categories of aphasia: a cluster analysis of Schuell test profiles. *Brit. J. of Disorders of Communications*, 111–122.

Prentice, H.C. (1979). Numerical analysis of intraspecific variation in European *Silene alba* and *S. dioica* (Caryoplyllocae). *Botanical Journal of the Linnean Society*, **78**, 181–212.

Prentice, H.C. (1980). Variation in *Silene dioica* (L) Clairv. Numerical analysis of populations from Scotland. *Watsonia*, **13**, 11–26.

Pritchard, N.M. and Anderson, A.J.B. (1971). Observations on the use of cluster analysis in botany with an ecological example. *J. Ecol.*, **59**, 727–747.

Rand, W.M. (1971). Objective criteria for the evaluation of clustering methods. *J. Amer. Statist. Assoc.*, **66**, 846–850.

Rao, C.R. (1952). *Advanced Statistical Methods in Biometrics Research*, John Wiley & Sons, New York.

Roeder, K. (1990). Density estimation with confidence sets exemplified by superclusters and voids in the galaxies. *J. Amer. Statist. Assoc.*, **85**, 617–624.

Rohlf, F.J. (1970). Adaptive hierarchical clustering schemes. *Syst. Zool.*, 58–82.

Rohlf, F.J. (1973). Hierarchical clustering using minimal spanning tree. *Computer J.*, **16**, 93–95.

Rohlf, F.J. (1974). Graphs implied by the Jardine–Sibson overlapping clustering methods, B_k. *J. Amer. Statist. Assoc.*, **69**, 705–710.

Rohlf, F.J. (1975). A new approach to the computation of the Jardine–Sibson B_k clusters. *Computer J*, **18**, 164–168.

Rohlf, F.J. and Fisher, D.R. (1968). Test for hierarchical structure in random data sets. *Systematic Zoology*, **17**, 407–412.

Rohlf, M.E. (1978). A probabilistic minimum spanning tree algorithm. *Inform. Process. Lett.*, **8**, 44–49.

Rubin, J. (1967). Optimal classification into groups: an approach for solving the taxonomy problem. *J. Theor. Biol.*, **15**, 103–144.

Saito, T. (1980). A hierarchical clustering method for rank order data. *Behaviourmetrika*, **8**, 23–29.

Sammon, J.W. (1969). A non-linear mapping for data structure analysis. *IEEE Trans. Computers*, C18, 401–409.

Scheibler, D. and Schneider, W. (1985). Monte Carlo test of the accuracy of cluster analysis algorithms—a comparison of hierarchical and nonhierarchical methods. *Multiv. Behav. Res.*, **20**, 283–304.

Schuell, H. (1965). *Differential diagnosis of aphasia*, University of Minnesota Press, Minneapolis.

Scott, A.J. and Symons, M.J. (1971). Clustering methods based on likelihood ratio criteria. *Biometrics*, **27**, 387–398.

Sibson, R. (1970). A model for taxonomy. *Math. Biosci.*, **6**, 405–430.

Sibson, R. (1972). Order invariant methods for data analysis (with discussion). *J. Roy. Statist. Soc.*, B, **34**, 311–349.

Sibson, R. (1973). SLINK. An optimally efficient algorithm for the single link method. *Computer J.*, **16**, 30–34.

Singleton, R.C. and Kautz, W. (1965). *Minimum squared error clustering algorithm.* Stanford Research Institute.

Skinner, H.A. (1978). *Dimensions and Clusters: A Hybrid Approach to Classification.* Alcoholism and Drug Addition Research Foundation, Toronto, Ontario, M5S 2S1.

Sneath, P.H.A. (1957). The application of computers to taxonomy. *J. Gen. Microbiol.*, **17**, 201–226.

Sneath, P.H.A. and Sokal, R.R. (1973). *Numerical Taxonomy*, W.H. Freeman & Co., San Francisco.

Sokal, R.R. and Sneath, P.H.A. (1963). *Principles of Numerical Taxonomy*, W.H. Freeman & Co., London.

Spath, H. (1980). *Cluster Analysis Algorithms*, Ellis Horwood Ltd., Chichester.

Spath, H. (1985). *Cluster Dissection and Analysis*, Ellis Horwood Ltd., Chichester.

Strauss, J.S., Bartko, J.J. and Carpenter, W.T. (1973). The use of clustering techniques for the classification of psychiatric patients. *Br. J. Psychiat.*, **122**, 531–540.

Symons, M.J. (1981). Clustering criteria and multivariate normal mixtures. *Biometrics*, **37**, 35–43.

Tan, W.Y. and Chang, W.C. (1972). Some comparisons of the method of moments and the method of maximum likelihood in estimating parameters of a mixture of two normal densities. *J. Amer. Statist Assoc.*, **67**, 702–708.

Thorndike, R.L. (1953). Who belongs in a family? *Psychometrika*, **18**, 267–276.

Titterington, D.M., Smith, A.F.M. and Makov, U.E. (1985). *Statistical Analysis of Finite Mixture Distributions*, John Wiley & Sons, New York.

Tukey, P.A. and Tukey, J.W. (1981). Preparation, pre-chosen sequence of views. In *Interpreting Multivariate Data* (V. Barnett, ed), John Wiley & Sons, Chichester.

Wallace, C.S. and Boulton, D.M. (1968). An information measure for classification. *Computer J.*, **11**, 185–194.

Ward, J.H. (1963). Hierarchical grouping to optimize an objective function. *J. Amer. Statist. Assoc.*, **58**, 236–244.

Wastell, D.G. and Gray, R. (1987). The numerical approach to classification: a medical application to develop a typology for facial pain. *Statistics in Medicine*, **6**, 137–164.

Wegman, E.J. (1972). Nonparametric probability density estimation. *Technometrics*, **14**, 533–546.

Wilkins, M. and MacNaughton-Smith, P. (1964). New prediction and classification methods in criminology. *J. Res. in Crime and Delinquency*, **1**, 19–32.

Wilkinson, L. (1992). Graphical displays. *Statistical Methods in Medical Research*, **1**, 3–25.

Williams, W.T. and Dale, M.B. (1965). Fundamental problems in numerical taxonomy. In *Advances in Botanical Research* (R.D. Preston, ed), 2, Academic Press, London.

Williams, W.T. and Lambert, J.M. (1959). Multivariate methods in plant ecology, 1. Association analysis in plant communities. *J. Ecol.*, **47**, 83–101.

Williams, W.T., Lambert, J.M. and Lance, G.N. (1966). Multivariate methods in plant ecology. V. Similarity analysis and information analysis. *J. Ecol.*, **54**, 427–445.

Williams, W.T., Lance, G.N., Dale, M.B. and Clifford, H.T. (1971). Controversy concerning the criteria for taxonometric strategies. *Comp. J.*, **14**, 162–165.

Wishart, D. (1969a). Mode analysis. In *Numerical Taxonomy* (A.J. Cole, ed), Academic Press, New York.

Wishart, D. (1969b). An algorithm for hierarchical classifications. *Biometrics*, **25**, 165–170.

Wishart, D. (1969c). Numerical classification method for deriving natural classes. *Nature*, **221**, 97–98.

Wishart, D. (1971). *A generalised approach to cluster analysis*. Part of Ph.D. Thesis, University of St. Andrews.

Wishart, D. (1973). *An improved multivariate mode-seeking cluster method*. Paper presented at Royal Statistical Society General Applications Section and Multivariate Study group Conference.

Wishart, D. (1978). *Clustan User Manual*, Program Library Unit, Edinburgh University.

Wolfe, J.H. (1969). Pattern clustering by multivariate mixture analysis. *Research Memorandum*, SRM 69-17, U.S. Naval Personnel Research Activity, San Diego.

Wolfe, J.H. (1970). Pattern Clustering by multivariate mixture analysis. *Multiv. Behav. Res.*, **5**, 329–350.

Wolfe, J.H. (1971). A Monte Carlo study of the sampling distribution of the likelihood ratio for mixtures of multinormal distributions. Naval Personnel and Training Research Laboratory. *Technical Bulletin*, STB 72-2 (San Diego, California 92152).

Wong, M.A. (1982). A hybrid clustering method for identifying high-density clusters. *J. Amer. Statist. Assoc.*, **77**, 841–847.

Wong, M.A. and Lane, T. (1983). A *k*th nearest neighbour clustering procedure. *J. Roy. Statist. Soc.* B, **45**, 362–368.

Wong, M.A. and Schaack, C. (1982). Using the *k*th nearest neighbour clustering procedure to determine the number of sub populations. *Proc. of the Statistical Computing Section, Americal Statistical Association.*

Zahn, C.T. (1971). Graph-theoretical methods for detecting and describing Gestalt clusters. *IEEE Trans. Computers*, C20, 68–86.

Zubin, J. (1938). A technique for measuring likemindedness. *J. Abnor. Soc. Psychol.*, **33**, 508–516.

Index